Table of contents

I0053530

Testing Protei_006 in Kaag, Netherlands

Introduction

Open_Sailing is a growing international community, with the goal of developing open-source technologies to explore, study, and preserve oceans. During the summer of 2011, a group converged in Rotterdam to develop Protei_006, the prototype for a potential fleet of low-cost, shape-shifting, DIY, semi-autonomous oil-collecting sailboats, that sail upwind and collect oil sheens on the water. This robotic sailboat is intended to be self-righting, durable, inexpensive, and easily reproducible.

The goal this summer was to develop, test and document the making of a 3-meter long Protei vessel in order to set up the primary steps for a comparison of efficiency between an articulated-hull boat sailing upwind and the existing pollution-cleaning technologies. Current methods for cleaning oil pollution on water, such as those implemented in the BP oil spill in the Gulf of Mexico, rely on expensive equipment and massive contingency plans, expose workers to toxic chemicals, and are constrained by weather conditions [Kerr, 2010].

Protei, in contrast, is unmanned, uses accessible, inexpensive or recycled materials, and tolerates chemically hazardous and rough weather conditions. Because of its open hardware and Do-It-Yourself philosophy, it can be constructed and deployed on a large scale in a cost-efficient way. It can eventually be appropriated for other purposes, such as cleaning other chemical pollutants and material waste off the water or ocean monitoring.

The design and implementation of Protei will be an ongoing evolution, as people all over the world continue to build and deploy it version after version, and contribute to its efficiency and robustness.

Solving a man-made problem with the power of nature

Transocean 2010

NASA 2010

On April 20th 2010, the Deepwater Horizon oil rig exploded, killing 11 men and injuring 17 others. For 3 months, 4.9 million barrels (780,000 m³) of crude oil gushed from the sea floor to create a 80-square-mile (210 km²) "kill zone" around the gusher. In addition of crude oil, 1,791,000 US gallons (6830 m³) of toxic oil dispersant were deployed on the surface and underwater.

The dispersant most widely used were Corexit EC9500A and EC9527A, both containing known carcinogens that damage animal and human reproductive systems and developing fetuses.

As the satellite image above shows, the spill was easily distinguishible from space but scientists also reported immense underwater plumes of dissolved oil not visible at the surface.

The BP oil spill is known to be the worst environmental catastrophe in the history of the North American continent and the largest accidental marine oil spill in the history of the petroleum industry.

The spill caused extensive damage to marine and wildlife habitats and to the Gulf's fishing and tourism industries. It will affect the health of the Gulf residents for a long time.

Skimmer ships, floating containment booms, anchored barriers, and sand-filled barricades along shorelines were used in an attempt to protect hundreds of miles of beaches, wetlands, and estuaries from the spreading oil. Overall, the technologies used prove to be inefficient while being very expensive, often exposing the health of clean up crew and vulnerable to difficult weather conditions during the hurricane season.

The map below shows a clear correlation between the surface currents (arrows) matching the extent of the oil spill (black) and the residents reporting oil spill (red dots).

The inefficiency of the cleanup effort was largely due to the lack of examination of natural patterns and local resident knowledge. [Anne Rolfes, 2010]

Protei was born from the hope that we may clean up an oil spill more efficiently with an appropriate technology that uses natural forces and pattern (wind, surface current) and local residents knowledge and creativity.

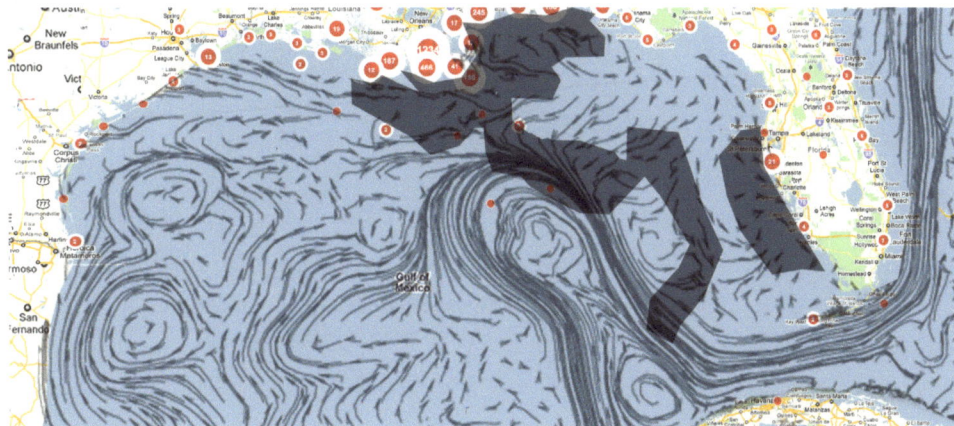

http://www.aoml.noaa.gov/phod/dhos/fig_model/JT/

http://rucool.marine.rutgers.edu/deepwater/

http://oilspill.labucketbrigade.org/

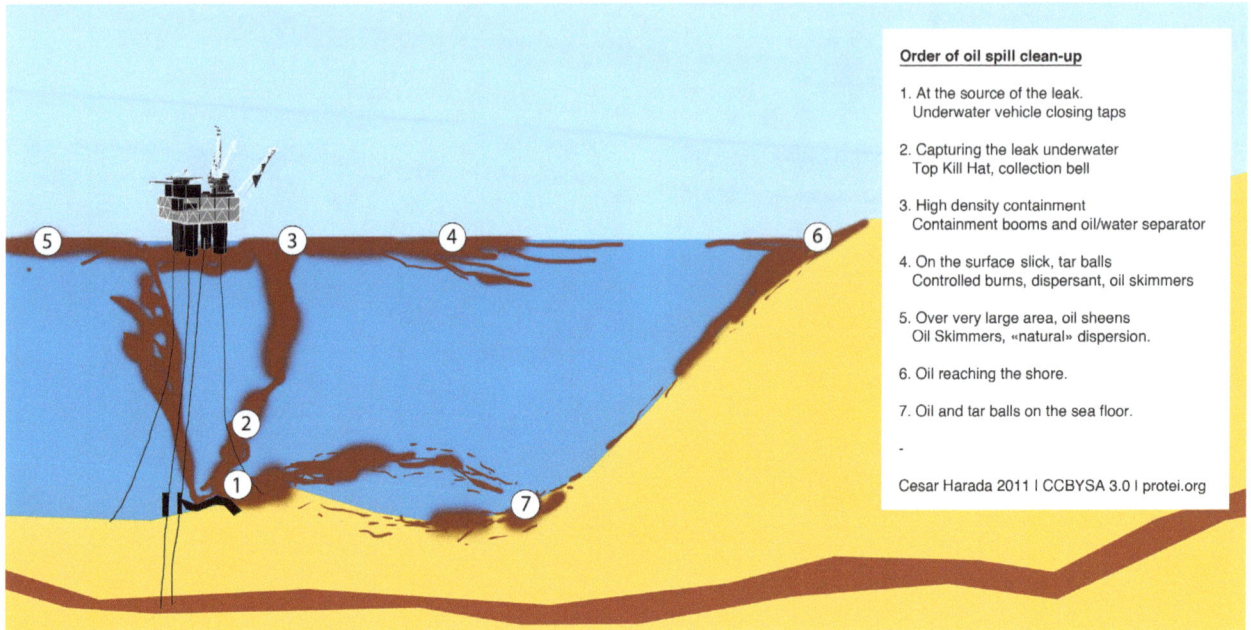

Protei research coordinator Cesar Harada, moved to the Gulf of Mexico during the oil spill and based on his observation, proposes a simplified classification of clean up efforts.

Protei would be most useful for cleaning oil spills of the fourth order, in which there is a thin sheen of oil that has spread throughout a large surface area of water, and that is travelling towards the land or the sea.

Protei would focus of collecting surface sheen of light oil.
Currently, oil spills of this nature and of this magnitude are often targeted by fishermen using repurposed fishing vessels. Fishermen use a combination of oil containment boom (orange) and oil sorbent boom (white) often made of recycled plastic. When the sorbent is saturated with oil, they are collected and treated.

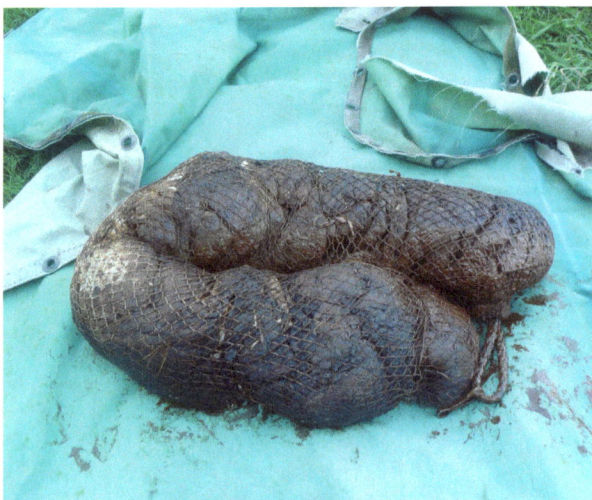

Oil boom saturated with oil

*Oil booms misplaced and entangled.
Barataria Bay, 2011-06-24*

The Protei approach

An oil spill is a man-made disaster but follows natural patterns to propagate: the wind and surface currents, forming long stretches of oil.

In the case of the BP oil spill, repurposed fishing vessels were used, pulling a combination of oil containment (orange) and oil sorbent (white). Fishermen were simply eyeballing oil from their boat exposing their health. They would not be able to work at night, in rough weather condition nor far from the shore. It is estimated that the hundreds of oil-powered skimmers collected only 3% of the oil spill.

With an equivalent length of oil sorbent pulled upwind, we may be able to collect a lot more oil.

If we multiply the rig, we may increase significantly the absorbtion performance. Now the problem is how to present this much surface area of oil sorbent against the powerful wind and surface currents.

What if we were sailing upwind, capturing the oil that is drifting downwind? As we are tacking, we would intercept oil in the successive folds of our sorbent boom. The idea is to deploy the maximum surface of oil sorbent over the largest area, using the minimum energy. Using the power of nature to remediate a man-made problem.

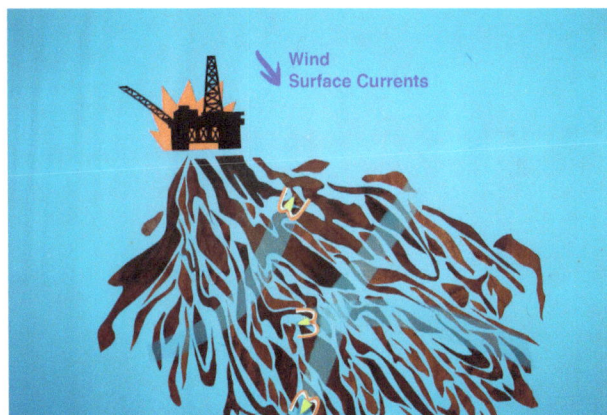

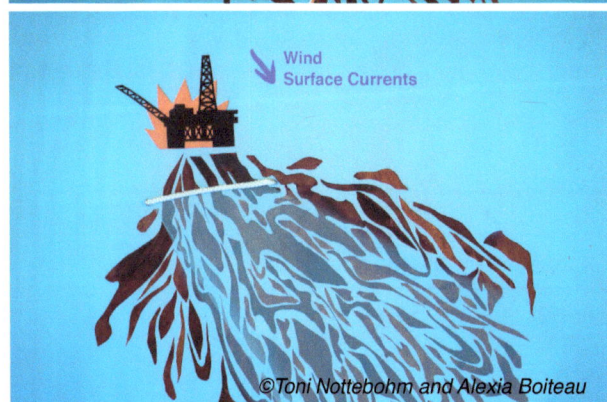

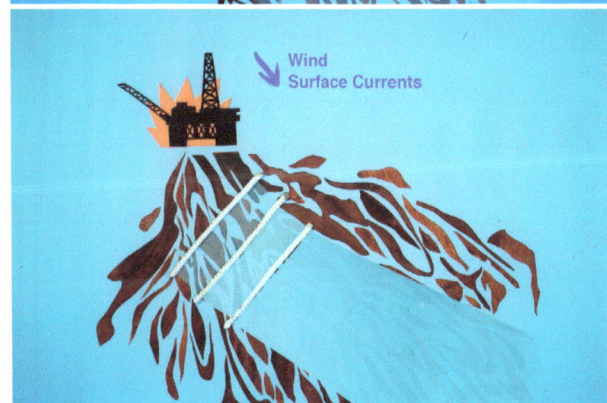

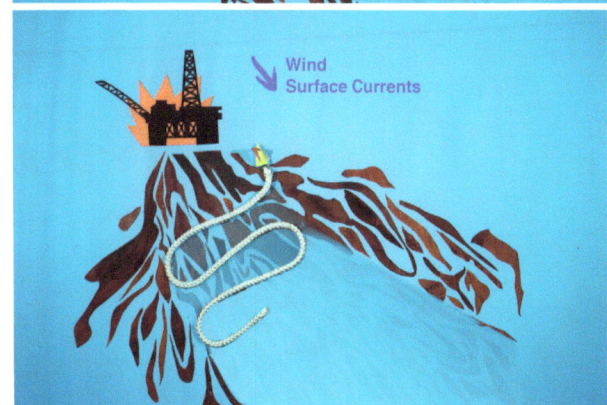

Wind
Surface Currents

©Toni Nottebohm and Alexia Boiteau

©Greenpeace

©AP

The following table demonstrates some of the benefits that Protei could provide over current methods of oil spill relief performed by fishermen and repurposed vessels.

REPURPOSED MANNED FISHING VESSELS	PROTEI GOALS
Exposes crew to health risks and toxins	Unmanned and autonomous
Cannot operate during a storm	Able to operate during extreme weather conditions
Sensing of oil limited to human eye sight	Sensing technologies on board coupled with oil spreading prediction models
Not sustainable and environmentally destructive, requiring oil or diesel for driving power	Uses renewable energy: solar power for the actuation and wind energy for locomotion
Expensive	Inexpensive
Proprietary design	Open-source hardware

©Greenpeace

Robotic ships to the rescue

Nearly one year after the Deepwater Horizon disaster — in which cleanup technologies could only collect 3% of the spill — the environmental organization **Open Sailing has developed an automated fleet of drones called Protei** that promises better results at lower cost. Moreover, its open-hardware policy means anyone is welcome to modify, produce, and distribute the design.

CURRENT SOLUTION

SPILL SOURCE

Boats zig-zag across oil slicks

DIRECTION OF WATER CURRENTS

ONLY 3% OF THE SPILL RECOVERED

IDEAL SOLUTION

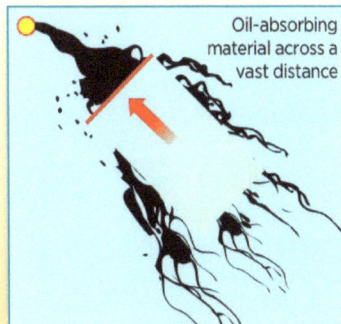

Oil-absorbing material across a vast distance

PROTEI

Snake-like movement allows multiple passes of same area

5 PROTOTYPES BUILT SO FAR

LARGE, LIGHTWEIGHT SAIL WITH GOOD PULLING POWER

ELECTRONIC SENSORS TO AVOID COLLISION, DETECT WIND DIRECTION AND POWER GENERATED

ABSORBS UP TO 25 TIMES ITS WEIGHT IN OIL

STEERING IN FRONT

Unlike most boats with the rudder in the back, Protei's rudder is in the front, and its flexible hull bends to turn, just like the movement of an animal.

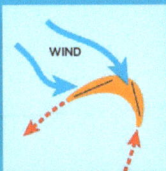

WIND

Open hardware: not owned by one company

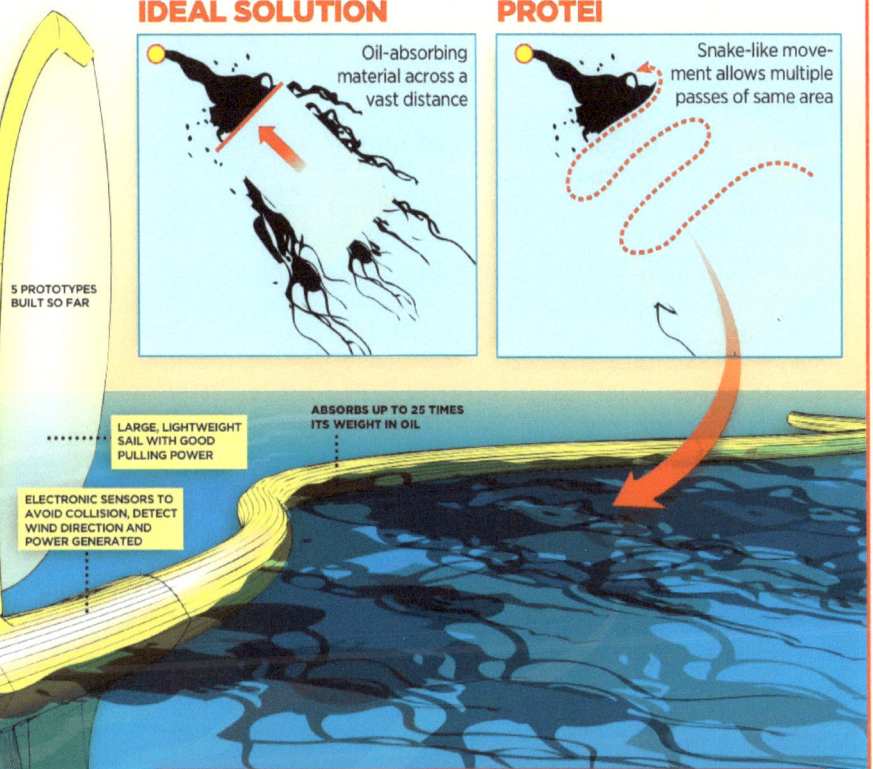

THE FLEXIBLE HULL ALLOWS THE BOAT TO HARNESS THE WIND'S POWER, EVEN WHEN TURNING DIRECTLY INTO IT. PROTEI NEVER LOSES THE PULLING POWER REQUIRED BY ITS LONG, HEAVY TAIL.

WHAT THE DESIGN MUST DO

■ Use existing technologies for rapid deployment

■ Sail semi-autonomously upwind, intercepting oil sheens going downwind

Must be:

■ hurricane-resistant
■ able to right itself if overturned
■ inflatable
■ unbreakable
■ cheap
■ easy to manufacture

ADVANTAGES

■ Unmanned, no human exposed to toxins.
■ Green and cheap, sailing upwind capturing oil downwind.
■ Able to operate in hurricane conditions.
■ Semi-autonomous : can swarm continuously, far from the coast.

NOT JUST FOR OIL SPILLS

The current design is meant for collecting oil, but it could be adapted to collect floating garbage, heavy metals in coastal areas, and toxic substances in urbanized waterways.

SOURCES : OPENSAILING.NET, PROTEI.ORG   RECHERCHE KINIA ADAMCZYK— INFOGRAPHIE JUSTIN STAHLMAN, AGENCE QMI

Protei & the science of sailing

Protei proposes the question: why not design a sailing boat with an articulated hull?
Many different appendages configurations can be thought of to steer a sailing boat:
1. "Classical sailing boat": stern rudder
2. Protei_001: bow rudder
3. Protei_005: bow and stern rudder
4. Protei_003 to _006: articulated hull — the whole hull is a rudder and a centerboard.

We would like to research and understand in the future what can be improved by using an articulated hull:

1) at low speeds

- Is the maneuvering capability improved when towing a long load in a zig-zag maneuver?
- Does articulating the hull help to tack?
- Is there less flow separation by having no appendages, hence less resistance?
- How to use the water flow to articulate the hull and use as little mechanical effort as possible to steer the boat, mimmicking fish locomotion?
- Can the force exerted on the rudder in a "classical sailing boat" be effectively distributed to the whole hull in Protei? Overall, does it require less steering energy to achieve the same maneuvering performance?
- Are the forces of the waves on the boat structure reduced due to the flexible body following the wave deformation of the free surface? For which wave lengths is it true? For which is it not?

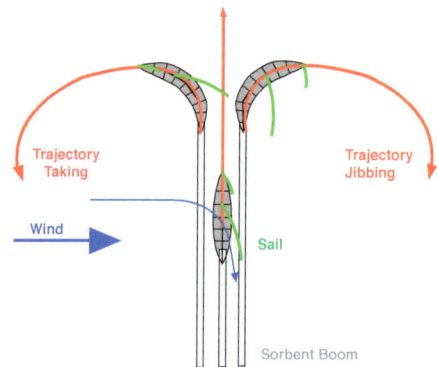

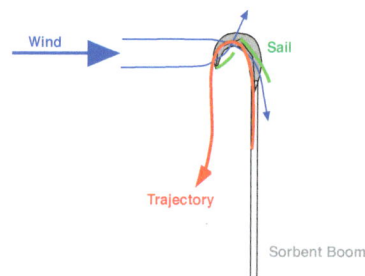

2) at high speeds

The effects of wave and flow become more prominent but the questions remain essentially the same.
In addition:

- Is it possible to model the articulated hull as a deformable lifting profile (for example a wing with flaps)? What does that tell us on how to articulate it best to improve the maneuvering capability?
- How to understand the interaction between the articulated hull and the sail trimming?
- Can we "trim" the shape of the hull in addition to trimming the sail to improve our speed, or maneuvering capability?
- How to trim both the hull and the sail then? What kind of rigging would interact best with this hull?

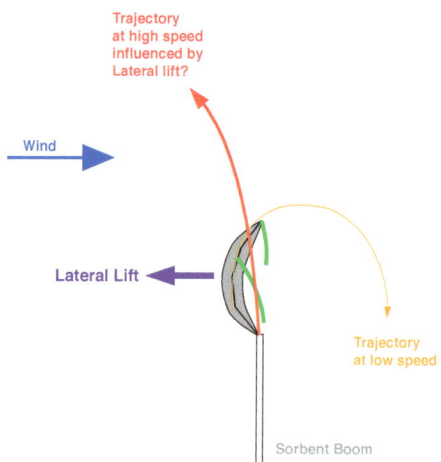

Trajectory Taking

Trajectory Jibbing

Wind

Sail

Sorbent Boom

Wind

Sail

Trajectory

Sorbent Boom

Trajectory at high speed influenced by Lateral lift?

Wind

Lateral Lift

Trajectory at low speed

Sorbent Boom

Scope

This handbook was written after the building of Protei_006 by the Protei_team between June 1st, 2011 and September 2nd, 2011 in Rotterdam. It is aimed at describing the design and construction of Protei. It is not a detailed construction handbook, but it should provide sufficient information to motivation, versions and research goals behind the design of the successive Protei versions.

The architecture of this handbook is highly modular so one can download and print on demand an isolated chapter and still have a self standing meaningful document. The content of this handbook is expandable, anybody is welcome to contribute to a section or suggest new ones. Just as we build successive versions of Protei, successive versions of this handbook are edited.

As well as providing background information about the project, this handbook includes an instructional section. This section is divided into the mechanical fabrication and the electrical design, as the Protei team was divided into these sections in building this prototype. However, the mechanical and the electrical elements should be developed in parallel, as they are closely linked.

Each section is divided into subsections corresponding to specific structural elements, for example, the sail, the winch, the cable system, and the GPS. For each element, there is overview information, specific information on its construction, and troubleshooting tips.

Intellectual Property

Protei is an open-source project. Its design is constantly evolving. Protei community members are willingly participating and sharing unique skillsets. This collaboration depends on accessible information and affordable components, as well as the sharing of ideas in a open environment.

Protei is made of open-source hardwares and softwares for the control, power and communication — between the vessel and an operator on land, including Arduino, Xbee, and a variety of sensors. Many of the mechanical components and materials are extracted from consumer and industrial products, such as power drills, plumbing equipment and salvaged wood.

The entire workflow of Protei is based on an open-source approach. Rather than a hierarchical chain of command, people share their own ideas for each aspect of the project, including planning, budgeting, hardware, software, fabrication, design, and testing.

In order to successfully grow into a fleet of DIY drones that can have impact on the oceans, the team documented, distributed and shared information about the mechanical fabrication and electronic architecture. This feeds back into Protei's success, which depends on ongoing contributions to its design and functionality.

The Protei team encourages the use of these materials, and of all materials that we have documented online and elsewhere, for the purposes of furthering the development of Protei as a concept and a technology. We hope that users will document and share any and all progress made using our material. Please make sure to cite Protei by providing a link to our URL, http://protei.org.

Open_Hardware

Protei's work is licensed under the following licenses, according to the type of content:

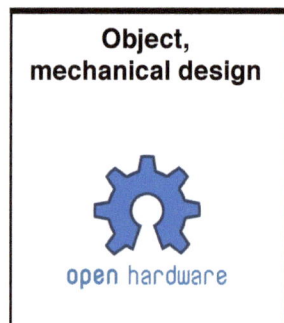

Object, mechanical design	Documentation, texts, photos, videos, communication materials	Source code	Name, trademark
open hardware	Creative Commons BY-SA	GNU General Public License, version 3 (GPL-3.0)	US trademark regulation #85339997

Open Source Hardware

Open source hardware is hardware whose design is made publicly available so that anyone can study, modify, distribute, make, and sell the design or hardware based on that design. Ideally, open source hardware uses readily-available components and materials, standard processes, open infrastructure, unrestricted content, and open-source design tools to maximize the ability of individuals to make and use hardware. Open source hardware gives people the freedom to control their technology while sharing knowledge and encouraging commerce through the open exchange of designs.

Creative Commons — BY-SA-3.0

This license lets others remix, tweak, and build upon Protei's work even for commercial purposes, as long as they credit Protei and license their new creations under the identical terms, by providing a link to our URL, http://protei.org. This license is often compared to "copyleft" free and open source software licenses. All new works based on ours will carry the same license, so any derivatives will also allow commercial use.

GNU General Public License 3.0

This license applies to the source code of Protei and requires that people have
- Freedom to run the program
- Freedom to access the code
- Freedom to redistribute the program to anyone (under the same license)

US Trademark #85339997

The name of PROTEI is registered under the serial number 85339997 by the United States Patent and Trademark Office.

Disclaimer

Protei is not the result of academic or scientific research. Protei is a direct response to environmental crises, including the financial and technological inaccessibility of solutions to the general public. It is an exploration of the concept of a flexible-hull sailing boat, and the consequences of this bio-inspired design. We understand that this design may not be the most efficient and that nothing we have done is completely new. Our vision is an artistic, collaborative, and open exploration of ultra-low cost, do-it-yourself sailing, and its application to oil-spill relief.

Protei_001

Before building Protei_001, we took a normal sailing boat out on the water, pulling a long heavy line. We observed that the longer the tail, the less pulling power and the less control on the direction. The initial question was: can we gain better control on a sailing boat by putting the rudder at the front? We just bought a simple wooden RC boat, modified it and observed that a front rudder offers better control and pulling capacity over a long heavy load.

Length: 100cm
Beam: 17cm
Height (total): 130cm
Displacement: 2.5kg
Material: wood, fiber-glass, resin, carbon tube
RC kit : Hitec Ranger 2ch
Boat kit: Tippecanoe T37
Cost : $300
Work time : 12 hours

With a 10cm rudder we were capable of towing and controlling a 400cm long sorbent behind Protei_001. That means using little energy to control a very large object.

Protei_001 is a modified RC wooden boat, the excellent Tippecanoe T37. Ordered for $265.00 on http://www.modelsailboat.com/t37sail.html, it arrived 3 days later, modified in a few hours.

Protei_001 was supported by the Louisiana Bucket Brigade and tested in the Pontchartain lake in New Orleans, LA.

Without the long sorbent behind, Protei_001 is a surprisingly fast and maneuverable RC Sailing boat. Since the rudder is at the front, one has a feeling similar to that of driving a car (which has steering at the front). The next step was thinking how we could gain even better control on direction and pulling power. The car analogy, and by extension the 4 wheel drive car - with steering both at the front and the back - gave us good hints of how we could add multiple points of directional control on the hull.

Protei_002

Based on the idea of multiplying the points of directional control on the hull, Protei_002 was developed. The initial idea was to have a rudder at the front, and another one at the back. In the design process, more than 2 points of control were envisioned, and the idea of an entirely articulated hull started.

Length: 100cm
Beam: 17cm
Height (total): 130cm
Displacement: 5kg
Material: wood, fiberglass, resin, PVC sheet and tubes
RC kit: Hitec Ranger 2ch
Boat kit: none
Cost: $200
Work time: 50 hours

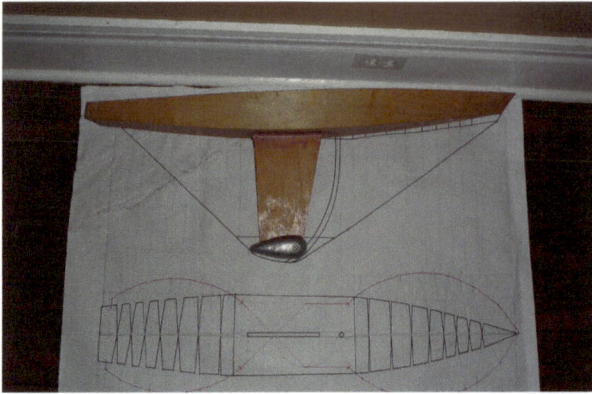

Protei_002 proportions are similar to Protei_001.
On the picture on the left, you can see Protei_001 on top of the technical drawing of Protei_002. The idea was to build 2 boats of similar sizes, one with nearly conventional characteristics and the other one with a revolutionary shape-shifting hull and actuation system.

You can see here Protei_001 (left) and Protei_002. Because Protei_001 is hollow, it is much lighter than Protei_002. The weight at the keel (ballast) was also tuned gradually, adding and removing weight.

Before installing the electonics in the hull of Protei_002, we did series of tests. We had 3 families of testing :
1. Protei straight (steady state)
2. Protei «S» shape test for stoping.
3. Protei «C» shape test for turning (picture).

We would pull the hull
1. by the «nose» (bow), by the «tail» (stern)
2. by the base of the mast.
3. by the top of the mast.

Eventually the remotely controlled tests were very successful. Protei_002 used the same RC kit as Protei_001, only the servomotor was replaced by a more powerful one (HS-765HB by Hitec). The main problem with Protei_002 was the weight: being to heavy, the waterline was too high and if the boat was to heel too much, water would get into the electronics and short-circuit the controls.
The next challenge would be to make a lighter structure.

Protei_003

Protei_003 was built to be an extremely lightweight structure.
Because of the size, the actuation would not be remotely done, but a man would be pulled behind Protei_003 in the water, simulating the drag a long heavy tail would create.

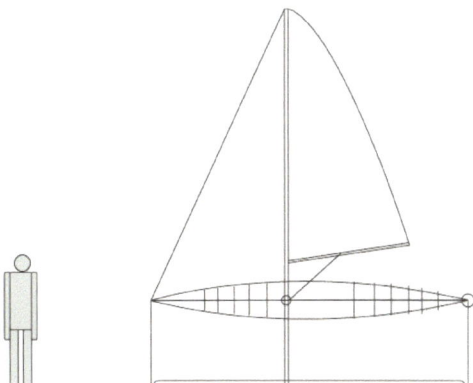

Length	400cm
Beam	120cm
Height (total)	600cm
Displacement	3kg + 40kg of ballast (sand bags).
Material	plastic sheet + duct tape. PVC tubes
RC kit	None
Cost	$150
Work time	80 hours

Protei_003 was built thanks to the support of Madam Suzette Toledano who graciously allowed us to use her garage just in front of the Pontchartrain lake.

Protei_003 was extremely lightweight and with its large sail surface it had great pulling power.

The command of the pilot that would be dragged behind Protei_006 was very simple: a bar with 2 lines to control the shape of the vessel, a self-locking line to trim the main sail. The pilot would have his body attached to the command bar.

Protei_003 was made of a recycled plastic that was very hard to weld or glue, so the air-tightness was insured by adhesive tape. Small air leaks occured and made the test inconsistent.

The area of the sail was great, and so was the ballast weight at the bottom. Protei_003 started to heel putting a lot of stress on the fragile connection between the mast and the keel, that eventually broke. The righting lever was too high. Great lesson learned, in the future we would connect the mast and the keel with a solid piece, or even better : make the mast and the keel of the same piece.

Building a large prototype with a very small budget was a challenge. It gave us the confidence to think of the future of Protei as a low cost technology.

Protei_005

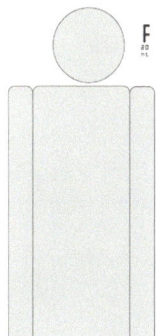

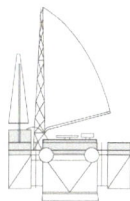

Protei_005 was designed to be as small as possible, yet containing much more advanced electronics: basic environmental sensing (wind, position) and collision avoidance logic.

Length	55cm
Width	28cm
Height (total)	80cm
Weight	2kg (including sand ballast).
Material	lunch boxes, copper rods, polystyrene float, tape Arduino, Xbee, ultrasonic Sensor, Wii Nunchuk
Cost	$150
Work time	150 hours

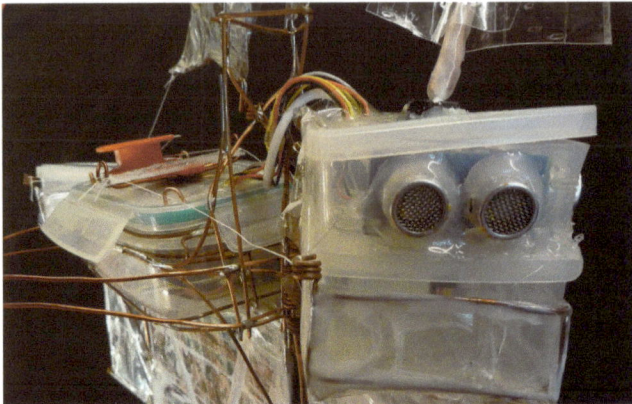

This version of Protei was built by Cesar Harada in collaboration with the randomwalks collective thanks to Nabi Center Seoul, Korea. On the left is shown the right ultra-sonic sensors informing Protei_005 about the promity of an obstacle so it could be avoided.

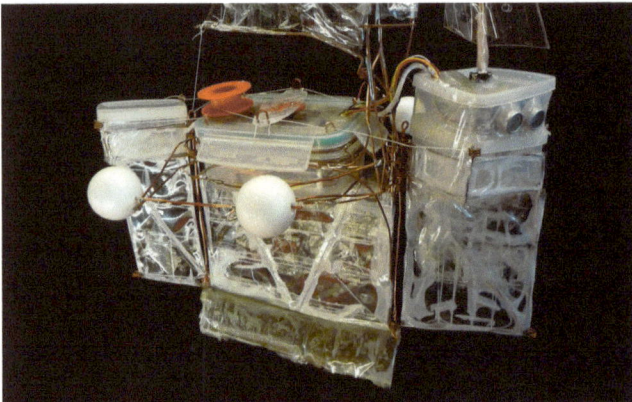

Protei_005 was so small, the position of the batteries, and the weight of the sensor threw it out of balance. We had to add polystyrene spheres to stabilize the vessel.

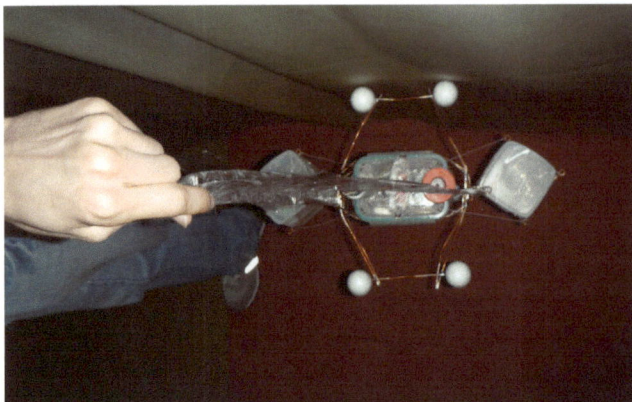

Protei_005 had a home-made wind sensor made from a hacked joystick, the Wii Nunchuk. The stripped joystick would have a "sail" on it to amplify the push of the wind. Protei_005 sensed and actuated its own shape efficiently out of the water.

During the test, Protei_005 leaned too much forward and the ultra-sonic sensors were pointing at the water, making the signal representing an obstacle very noisy. Protei_006 was shaking left to right and we had difficulties "cleaning up" our environmental sensing inputs.

Protei_005 established a good base for the electronic architecture using open hardware electronics and hacked consumer goods.

Protei_005.1

Protei_005.1 was designed as an enhanced version of Protei_002, inflatable and articulated, with a simple RC kit for all the controls. This larger version was designed to test the strength of the RC servos.

Length	130cm
Beam	40cm
Height (total)	160cm
Displacement	8kg (including brick ballast).
Material	lunch box, heat-sealed vaccum bags, plastic sheets, rope, pvc tubes.
Cost	$100
Work time	60 hours

Protei_005.1 was built at the V2_ Institute for the Unstable Media in Rotterdam, NL.

The body was built from vaccum bags, heat-sealed with teflon plyers. The balloons built stayed for 3 months with constant internal pressure. We got surprisingly good results for this domestic product.

The motors and batteries of Protei_005.1 were not strong enough to actuate the shape or to trim the sail.

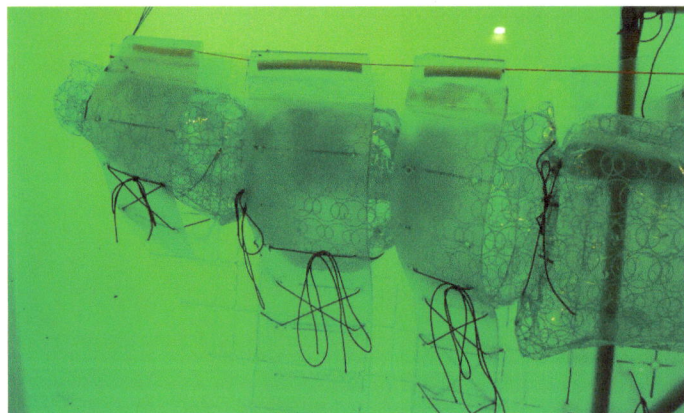

Even if it did not navigate with the radio control, it was a good test to learn about materials and fabrication of small prototypes. It also informed us that a regular RC kit is not powerful enough for this scale of model.

Working with ropes, we could make an inflatable that is adjustable in size. That would also be very useful for future prototypes, giving us an object that is easy to tune and modify. That is even faster than versionning.

We did sail Protei_005.1 passively in the Maas river in Rotterdam. With a fixed trim of the sail, Protei's inflatable body followed the wave motion.

Protei_005.3

After testing Protei_005.3 in the harbor by the studio, Rotterdam, June, 2011

Protei_005.3 is an enhanced version of Protei_002. The hull is mostly made of foam sandwiching one layer of wood. It is divided into 3 equal sections of 38 cm each.

Length	108cm
Beam	12cm
Height (total)	160cm
Displacement	6kg (including sand ballast).
Material	Plywood, foam, plastic
RC Kit	Futaba 2ch, 40Mhz
Servo	Regatta winch, Carson CS-13 digital servo
Cost	$400
Work time	70 hours

The segmented, articulated bow of Protei_005.3

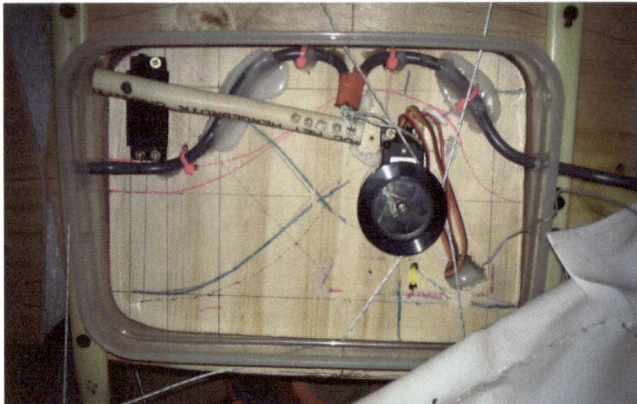
The electronics compartment

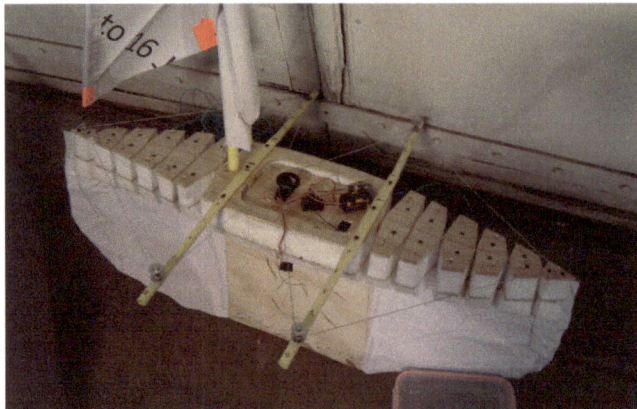
The hull and keel; The outriggers.

Side view prior to testing

This version has two motors, one which is a servo with an extended arm to trim the sail, and the other which reels a winch attached to two lines that control the articulation of the bow and the stern simultaneously.

The bow and stern section are fully segmented, and the middle section houses the electronics.

Because of the high torque required to reel the winch to articulate the hull, there are outriggers on either side to decrease the angle and reduce the torque.

Protei_006

Introduction

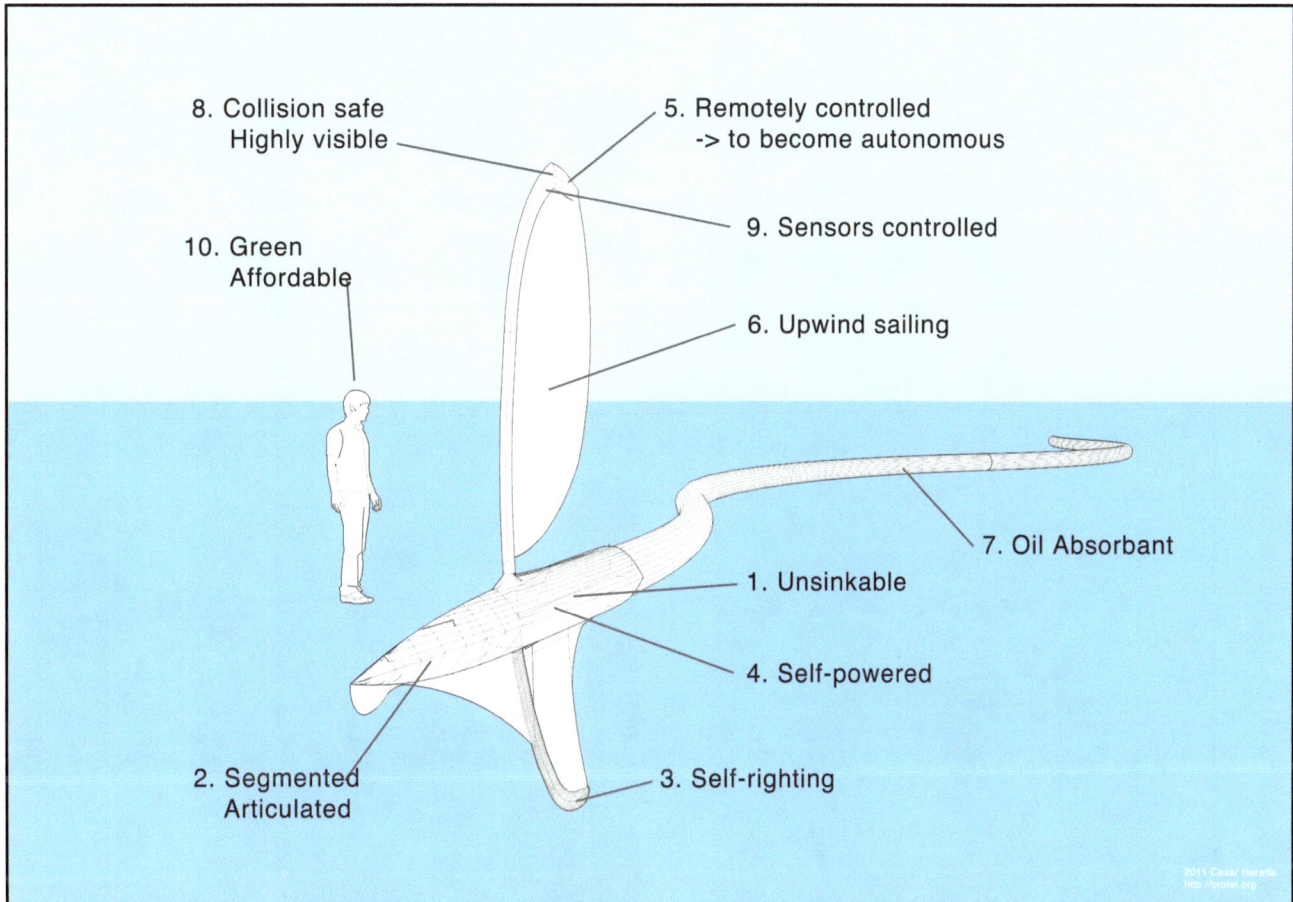

8. Collision safe
Highly visible

5. Remotely controlled
-> to become autonomous

9. Sensors controlled

10. Green
Affordable

6. Upwind sailing

7. Oil Absorbant

1. Unsinkable

4. Self-powered

2. Segmented
Articulated

3. Self-righting

Protei and its criteria: Unsinkable, segmented / articulated, self-righting, self-powered, remotely controlled and developing energetic autonomy, upwind sailing, highly visible / collision safe, green & affordable.

General Specifications

Length overall	: 3.0 m
Beam	: 0.42 m
Mast height	: 3.78 m
Max draft	: 1.47 m
Voltage	: 14.4 V
Max current draw	: 45 A
Displacement	: 120 kg
Desired steady state speed	: 2 knots

Protei Handbook | v.2011 09 02 | CC BY-SA-3.0 | contact@protei.org | p. 30 / 99

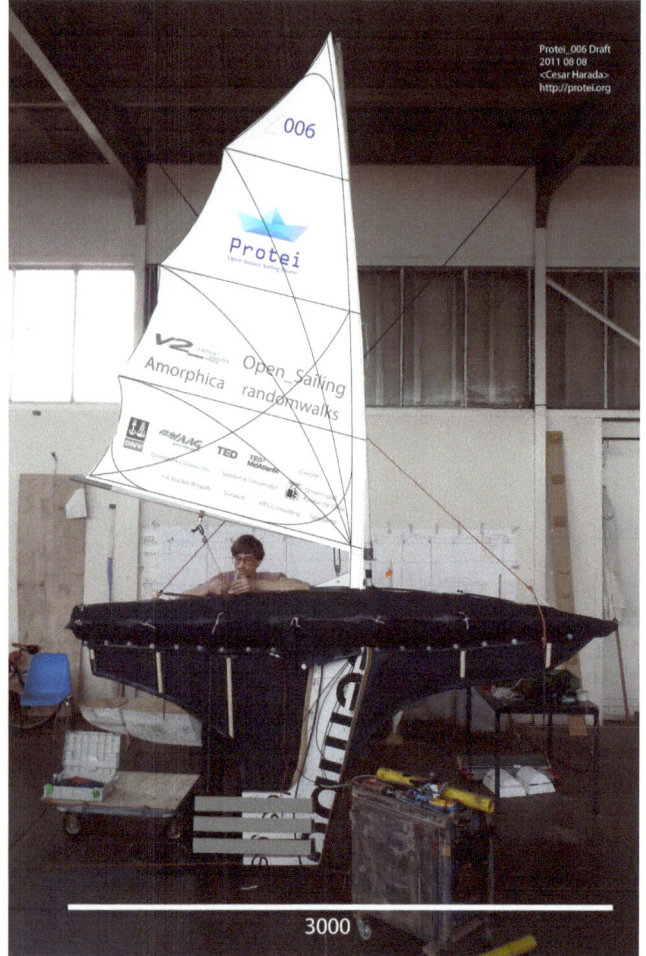

System Overview

Protei_006 is a 3m remote-controlled boat, controlled over radio by a user within 500m and line of sight. It has a segmented, shape-shifting hull, constructed of flexible spines that run lengthwise (through cross sectional bulkheads) and bend under stress. A flexible EPDM-Foam skin encases the skeleton. There is one 4m tall sail, and a large keel that extends downwards 1.2m from the hull. The majority of the boat's ballast is housed at the lowest point of the keel, which provides most of the stability. This also enables Protei_006 to be self-righting.

For steering and control, Protei_006 has three motors. The first one spins a winch to control the trim of the sail. The other two are attached to linear actuators that each control the articulation of the bow and the stern, which curve independently. The actuators pull and release cables that run throughout the longitudinal spines of the hull.

Protei_006's body is nearly cylindrical and very unstable. Most righting momentum comes from the ballast to right the vessel.

The battery and linear actuators, which provide most of the weight of the boat, are at the bottom of the keel, along with a 25kg lead ballast. The main electronics, the winch, and the GPS are housed in the waterproof, shockproof pelican cases[1] compartments of the hull (which make up the separate segments). Protei_006 stores GPS data from its trips.

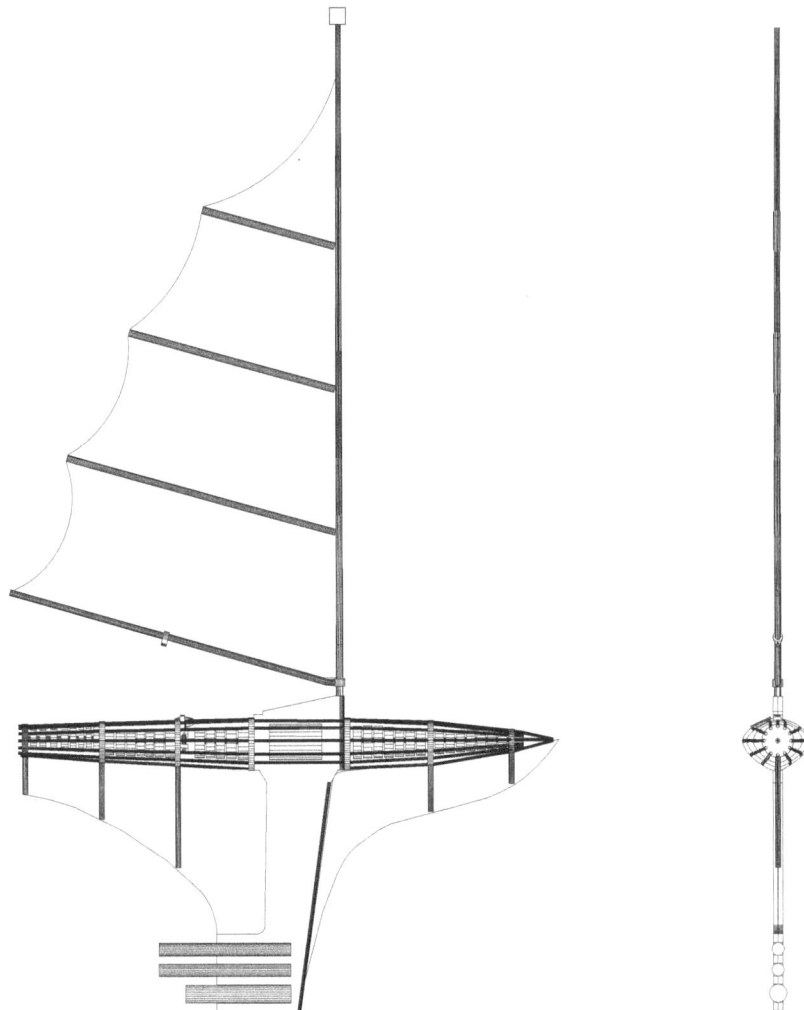

---

1   http://www.pelican.com

Goals

In order to make suitable technical choices for the materials and methods used in the construction of Protei, we have listed the global properties that we want to achieve. We implemented a solution for each property. The criteria are meant as refinements to achieve each property, which can be associated with a testing protocol, used to validate the technical solution.

For example, the criteria that Protei_006 be self-righting is tested by manually heeling the boat in the water to 90° and evaluating if it returns its original position. If it does so, we can state that the combination of the ballasted keel and the light superstructure meets the self-righting criteria.

Property	Criteria	Solution
Maneuverable	Steerable while towing	Shape-shifting through a flexible hull
	Optimized to sail upwind	Control of the sail
Stable	Self righting	Ballasted keel
		Light superstructure
	Unsinkable	Puncture-resistant, buoyant material
		Ample flotation
	Robust	Neoprene skin and use of sturdy materials
Safe	Collision-safe	COLREGS-compliant lighting system
Towing capable	Can tow sorbent	Enough pulling force Boom hooking point
Unmanned and automated	RC (for steering and sail trim) up to 1000m)	Xbee module
	Determines position data	GPS Module
	Store GPS data	SD card shield
	Power	Battery

Naval Architecture

Length Overall	Loa	3042	mm
Length Waterline	Lwl	3023	mm
Breadth Max	Bm	422.9	mm
Breadth Waterline	Bwl	419	mm
Draft Design	Td	1473	mm
Draft Canoe	Tc	161	mm
Design Displacement Volume	Disp.d	0.10	m³
Design Weight	W.d	105	kg
Waterplane Area	Awp	0.94	m²
Midship Section Are	Ams	0.05	m²
Coefficient of Fineness	Cw	0.74	—
Midship Coefficient	Cm	0.70	—
Block Coefficient	Cb	0.59	—
Prismatic Coefficient	Cp	0.84	—

A technical overview of Protei's mechanisms:

Propulsion	Wind propelled Sailing vessel, designed for up-wind sailing
Steerage	Flexible keel / rudder running length of hull Motor propelled 2x Bosch hand-drill motors
Power supply	14.4V, 10Ah Nickel metal hydride battery
Sail trimming	Motor controlled winch for sail trimming 12V windshield wiper motor
Crew	0

Weight Distribution

Taking all the components of Protei into account, a detailed weight distribution is constructed with corresponding centers of gravity for each part. It is important to gain a good understanding of any vessel's overall center of gravity for use in stability analysis and to make an estimation into her response to the elements she will be exposed to.

The following table represents the total weight for each part of the Protei_006 construction. The keel consisting of the wooden central keel and 'neoprene fins' acts as a rudder. The hull includes plywood ribs, tubular modules including wires, neoprene skin and instrumentation used for controlling and moving the sail. Nuts, bolts, screws, washers, and small parts are taken into consideration as are the 'additional parts' comprising the ballast and motors. Finally, the new mast, rigging and sail data are added to produce the total calculated weight equal to 105 kg.

Part / Description	Weight	TCG	Lever	VCG	Lever	LCG	Lever
-	(kg)	(mm)	(Mx)	(mm)	(Mz)	(mm)	(My)
KEEL	24.00	0.00	0.00	-390.27	-9367.08	1502.28	36057.23
HULL	42.46	-14.46	-613.82	-22.02	-935.03	1305.99	55454.04
MAST & RIGGING	5.64	0.00	0.00	1157.46	6528.09	1260.02	7106.54
ADDITIONAL PARTS	33.10	0.00	0.00	-1354.03	-44818.50	1293.00	42798.30
TOTAL	105.20	-5.83	-613.82	-461.89	-48592.51	1344.22	141416.12

It is important to remember that Protei voluntarily takes-on water! The hull, although designed with a skin, allows for water ingress and this is not included in the calculation, despite it being an area of concern for the weight and displacement calculations.

The weight calculated above corresponds to what will be referred to as the
Calculated displacement volume   = 0.10 m³

A comparison of this displacement is made with the underwater volume displacement calculated according to the volumes of the associated parts as shown in the table below:

Part / Description	Volume	Density	Weight
-	(m³)	(kg·m⁻³)	(kg)
HULL (underwater body)	0.0892	1025.00	91.457
Wooden KEEL	0.0256	1025.00	26.257
Neoprene KEEL	0.0053	1025.00	5.421
		total	123.135

The total weight, equal to 123kg, corresponds to a displacement in sea water ;
Design Displacement Volume   = 0.12 m³

The difference between the calculated and design displacements is thought to be due to the water taken on-board.

Centers of Gravity
From the weight distribution calculation, Protei's center of gravity is estimated as :

TCG (transverse)   = -5.83mm
Insignificant offset to starboard

VCG (vertical)   = -461mm from central axis, not WL

Positioning of ballast at keel base has lowered the CoG, however this might not be enough for adequate stability.

LCG (longitudinal)   = 1344mm from transom

Further analysis is recommended to position the LCB and estimate the trim angle.

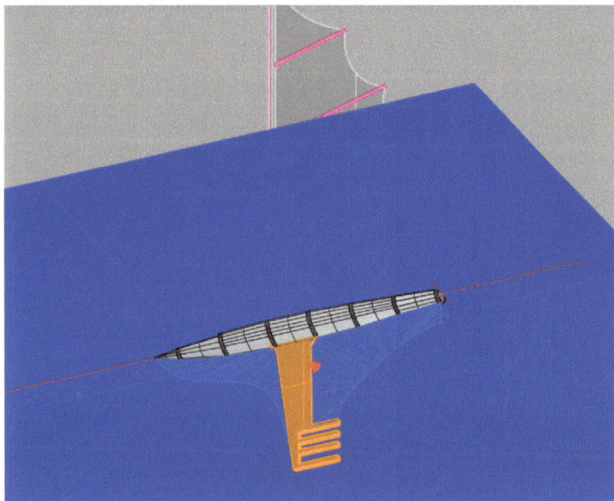

Protei below waterline, body and keel. The center of gravity is Illustrated by a red sphere.

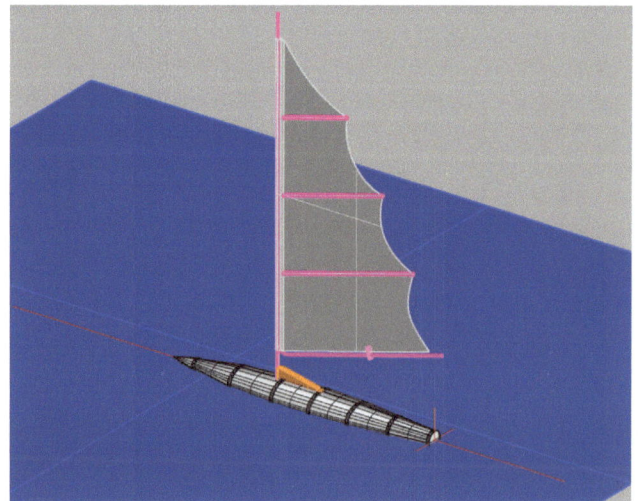

Protei above waterline, profile of body and sail

Protei dimensions

The following three images illustrate the main dimensions of Protei with respect to its above and below waterline body. These are summarised in the 'Protei Specifications' and used throughout the calculations.

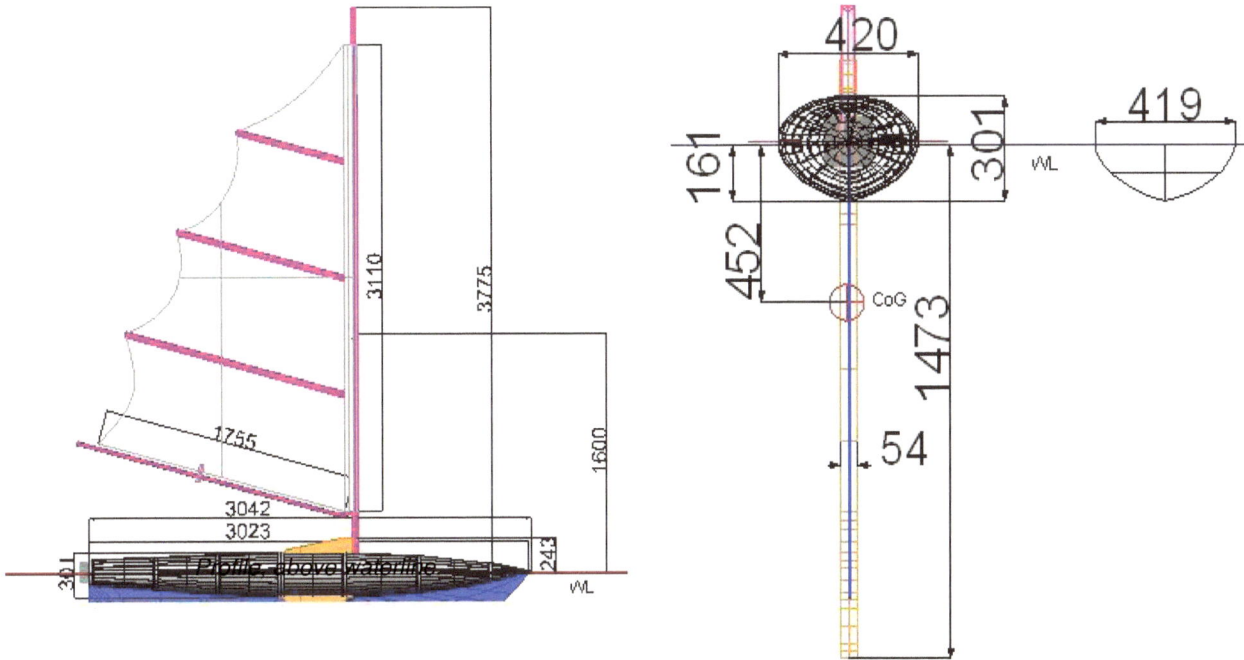

Side view with hull dimensions. The center of gravity is illustrated

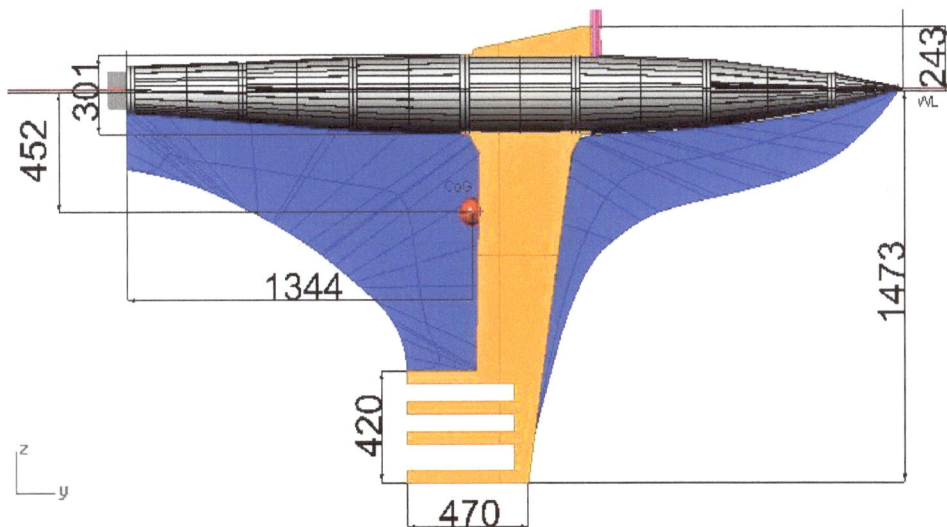

Profile, below waterline, with body and keel dimensions
The center of gravity is illustrated.

Stability

The stability of a yacht is generally governed by its keel, the ballast weight / bulb weight and the righting moment it creates to counteract the side force produced by the wind acting on the sail.

In large ships, the center of buoyancy (CoB) is generally positioned below the center of gravity (CoG). When heeled, the moment to correct the angle of heel, known as the 'righting moment', is created by a shift in its center of buoyancy. In the case of Protei though, its CoG is below her CoB and her almost cylindrical body doesn't have the ability to create a corrective buoyancy force when heeled. With a large amount of ballast positioned in her keel however, a righting moment is created and it is this factor, that has been investigated so far. The sail data for this investigation is detailed below.

Sail area	3.01 m²
Sail center of effort (z plane) from WL	1600 mm
Luff length	3110 mm
Foot length	1755 mm
Aspect ratio	3.54
Side force coefficient	1.25
Driving force coefficient	0.315

Both the driving and heeling forces created by the wind on the sail at various windspeeds can be calculated using this data. The driving force shall be used to propel Protei through the water whilst the heeling force would be counteracted. The latter should be accomplished by the keel and her ballast weights. A typical wind speed that Protei may be exposed to is 20 knots as highlighted in the table below detailing the moments and forces created by the wind.

Wind Speed	Wind Speed	Side Force	Side Moment	Side Moment	Driving Force	Driving Moment	Driving Moment
knots	m·s⁻¹	N	N·m	kg·m	N	N·m	kg·m
0	0	0	0	0	0	0	0
5	2.57	14.74	23.58	2.40	3.71	5.94	0.61
10	5.14	58.94	94.30	9.61	14.85	23.76	2.42
15	7.72	132.62	212.18	21.63	33.42	53.47	5.45
20	10.29	235.76	377.22	38.45	59.41	95.06	9.69
25	12.86	368.38	589.40	60.08	92.83	148.53	15.14
30	15.43	530.46	848.74	86.52	133.68	213.88	21.80
35	18.00	722.02	1155.23	117.76	181.95	291.12	29.68
40	20.58	943.04	1508.87	153.81	237.65	380.24	38.76

From 20 knots, 38 kg·m of side-force is applied to Protei. A crude analysis method has been adopted to form these results, but a guideline is nevertheless produced. The angle of heel corresponding to the righting moment required to counteract this force is predicted to be around 55 degrees. This can be observed in the table on the next page, showing the relationship between heel angle and righting.

Heel Angle	Lever	Righting Moment	Righting Moment
deg	mm	kg·m	kg·mm
0	0	0.00	0
5	39	4.14	4143
10	78	8.26	8255
15	117	12.30	12304
20	155	16.26	16260
25	191	20.09	20091
30	226	23.77	23770
35	259	27.27	27268
40	290	30.56	30558
45	320	33.62	33616
50	346	36.42	36418
55	370	38.94	38943
60	391	41.17	41171
65	410	43.09	43086
70	425	44.67	44673
75	436	45.92	45920
80	445	46.82	46818
85	450	47.36	47359
90	452	47.54	47540

The calculations are the basis of a static calculation assuming constant wind conditions. Waves and current have not been taken into consideration. A number of factors such as these can be analysed in more detail when a more thorough investigation is carried out. These could include the effect of side force on the sail in relation to its changing angle of heel which in this case is assumed constant (it is therefore likely to lessen in reality), and the angle of attack in relation to the position of sail and sail trimming, which will effect the sail coefficients.

Nevertheless, the following study of windspeed could be used to modify the future design. The figure below shows the righting moment against angle of heel.

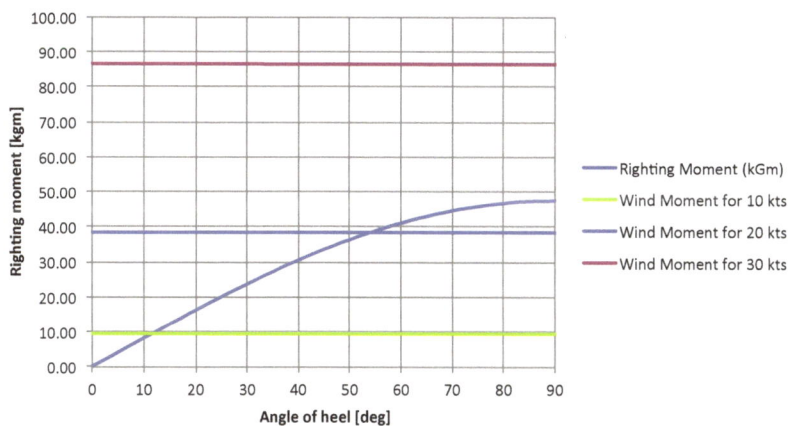

Legend:
- Righting Moment (kGm)
- Wind Moment for 10 kts
- Wind Moment for 20 kts
- Wind Moment for 30 kts

From the graph displayed, it is concluded that Protei will no longer be able to self-right when exposed to winds above 22 knots, > Beaufort 5. In order to increase her performance across a greater range of wind conditions a few options can be considered;

- Increase her ballast weight further
  » limited by her buoyancy / displacement
- Increase her draft
  » again limited by her buoyancy, her ability to be handelled and operable area

Considering Beaufort 5 as the top end of sailing ability, however, is not realistic. This value will be downgraded due to the dynamic performance of Protei and the fact that waves will be present in a true situation. As a topic requiring further investigation, it is perhaps one option to consider that at this stage, the most efficient method of improving her operability range is to up-size Protei.

Discussion of performance

Protei's design is unique. Although not all factors and goals are manageable at this stage, the innovative ideas spark discussion and further developments in the design to match the challenging specifications. The performance of the vessel is vital for her role, therefore the characteristics determining her performance have to be considered carefully.

Hull Design

On viewing Protei for the first time, the impression is that its profile is too cylindrical and it has too little breadth, meaning that ultimately it is likely to heel and capsize very easily. It is not a conventional sailing yacht although the principles of sailing can still be applied; it is after-all a vessel designed to work with, not against, the elements.

The driving force created by the wind on the sail is not considered in detail at this stage in design, although it can be noted that Protei's slender hull lines are designed to optimise her passage through the water by creating minimal drag. True optimisation is yet to be undertaken, however, a long thin hull shape generally travels at greater speeds and with less resistance than a broad, short hull form, therefore it can be considered a good starting point. When the task is adopted however, attention should be given to the rudder material and shape in order to ensure that minimal drag is created by its distortion when underway. Additionally, a study using a VPP (Velocity Prediction Program) and the driving force from the sail can be undertaken.

The Elements

It is in the design requirements that Protei shall be able to cope with extreme conditions; namely extreme wind and waves acting on the hull and sail in the form of extreme gusts and storms. These conditions could cause damage and even destruction to any part of the vessel. To avoid such an outcome Protei is stated to have a 'safety' procedure which involves her in a position of 'full-heel' — heel over by 90 degrees — with the sail flat on the water.

For a sailing vessel to survive extreme conditions in this position is very unlikely.

A number of questions are also raised as to how Protei would recover from this scenario should it come about. With factors such as the sail filling with water, the waves crashing over the hull, the entanglement of the tail and potential damage to the structure encountered, to consider, this self-righting / recovery characteristic makes a good foundation for investigation. The probability and the limits to which it could survive should be incorporated into such an investigation, especially if a more robust prototype is constructed.

Navigating through Oil

However harsh the weather elements are, Protei also has an extra environmental factor to consider. Navigating through oil will not only be slow, it will be severely hindering to the structure, its components and its ability to maneuver. Due to its density and viscosity, the oil in and on the hull will increase its weight, hence draft of the vessel, finally affecting its stability, maneuverability and overall performance.

Maneuverability

Protei has a combined rudder and keel. This combination is used as a mechanism for maintaining its course and turning, achieved by the bending of the hull and underwater body. The moment created by this bending is designed to turn the whole vessel. There are a number of concerns that arise from this type of steerage some of which include:

- Whilst underway, will Protei be able to maintain a heading?
  - » A normal yacht or dingy requires constant adjusting or holding in a position (off centre) of the helm / rudder in order to keep the vessel on course.
  - » Maintaining a heading will not be a matter of continuing close hauled until requiring a tack. The idea is to have it constantly adjusting its position of sail – ideally between broad-reach and close-hauled (averaging a close-reach).
  - » This is to be achieved by the bending of it hull creating the force to turn it into or away from the wind. Precise and quick control is required to eliminate the risk of travelling to far down or up-wind trying to avoid any risk of stalling or gybing.
  - » Only possible if the mechanism is powerful and quick enough to react to the conditions, (including gusts).
- Will the turning moment created be strong enough to tack?
  - » Protei has to pass through the waves / against the current and through the wind to fulfill a tacking manoeuvre.
  - » At this stage this has to be monitored during trials.
- How much will the tail influence the steering and turning ability of the hull?
  - » It is feared that when the tail is installed on Protei her ability to maneuver will be lessened as might its ability to tack.

Observation and sea trials can be used to verify such queries at this stage in the design.

Rolling

Protei's cylindrical body is unlikely to be able to maintain a constant angle of heel. Exposed to the inconsistencies of the wind, the attempt to find a position of equilibrium will be ongoing. Due to the righting moment being determined by the offset of CoG from the centreline, it is likely that it adopts a rocking motion whilst underway. With no crew on board to consider, there is no danger of creating an uncomfortable environment. What will be a concern, however, are the losses felt when energy that could be used to propel Protei is used to correct her sailing position.

Although Protei is faced with meeting extremely challenging objectives, the ideas and investigations, ongoing and future, are promising. With open discussions and open minds it is possible to meet a few more of these objectives in the future.

Preparing Protei_006 for launch

Mechanics

System overview

Protei_006 is the first large scale Protei prototype. It is designed to be an articulated sailboat that can eventually be constructed to carry equipment that absorbs oil and conducts remote sensing procedures at sea. The structure has a bio-inspired design, most significantly represented in its cylindrical hull profile. The construction of Protei is meant to be replicable by amateur builders and enthusiast. This section describes the process of building and assembling the mechanical components of Protei_006. In addition to this document, CAD files and drawings are provided on the Protei's git repository for laser cutting and CNC machining:
https://github.com/Protei/Protei-005-6

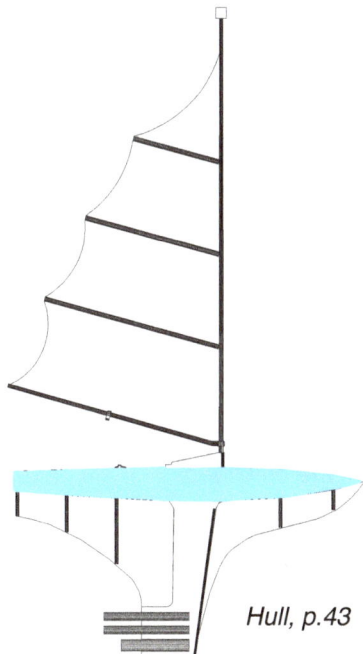

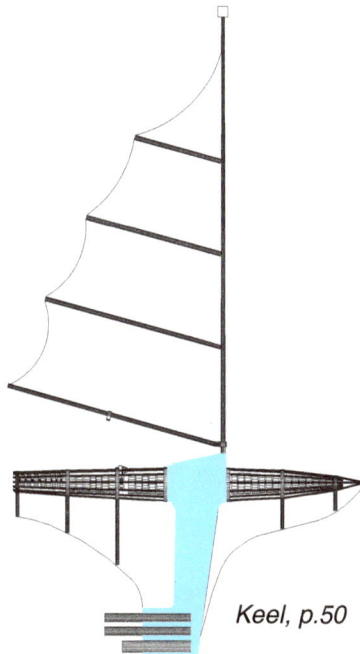

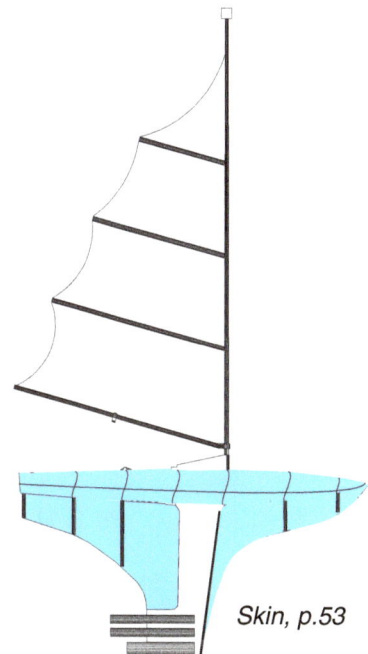

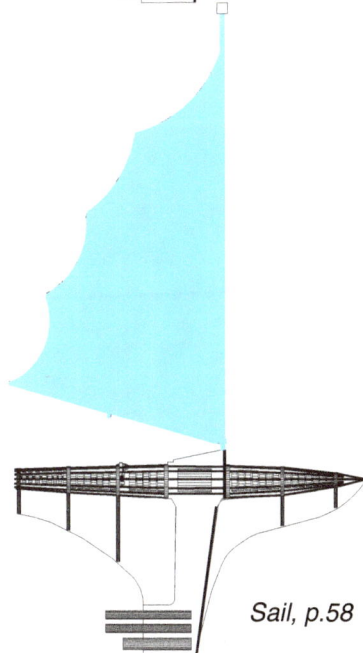

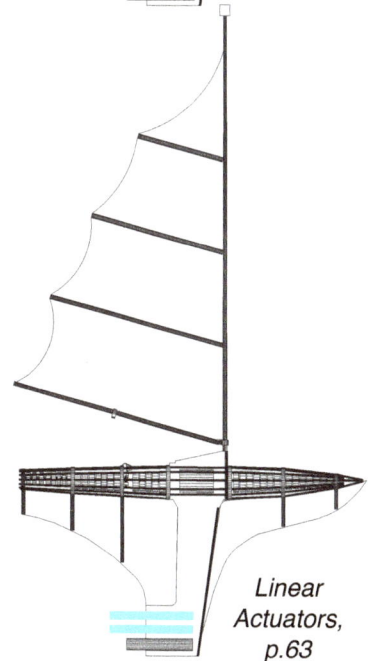

Hull, p.43

Keel, p.50

Skin, p.53

Sail, p.58

Winch, p.60

Linear Actuators, p.63

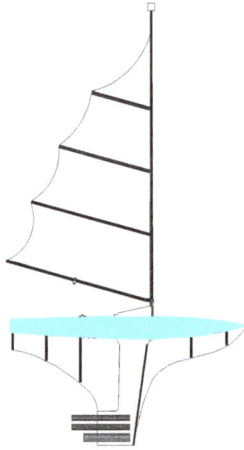

Hull

The hull of Protei is based on a skeletal structure composed of cross sectional bulkheads and longitudinal spines. The hull is divided into segments that are divided by bulkheads (sometimes referred to as "ribs") which define the shape of Protei_006. The bulkheads are connected via PVC tubes that are referred to as "spines". There are twelve outer spines that define the shape of the hull and one central spine that holds the bulkheads together around which everything bends. The spines are comprised of tubes of different diameters so that they slide into each other and allow the hull to bend. The outer spines are attached to each bulkhead by rivets, which allows each segment to be dismantled while still connected to its spines.

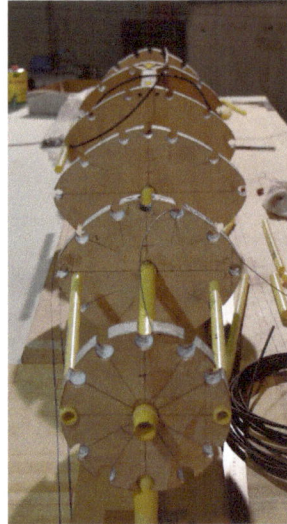

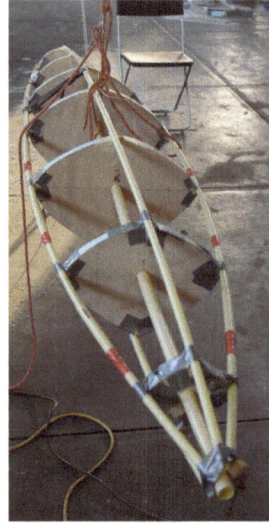

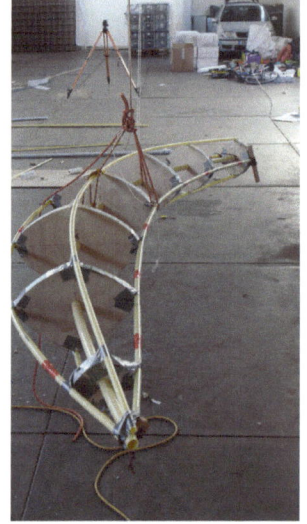

Hull construction and assembly of the bulkheads

Hull Construction and Assembly

The hull construction and assembly process can be divided as follows:

A) Bulkhead Fabrication
B) Pipe Cutting and Assembly
C) Riveting the Hull
D) Adding Flotation

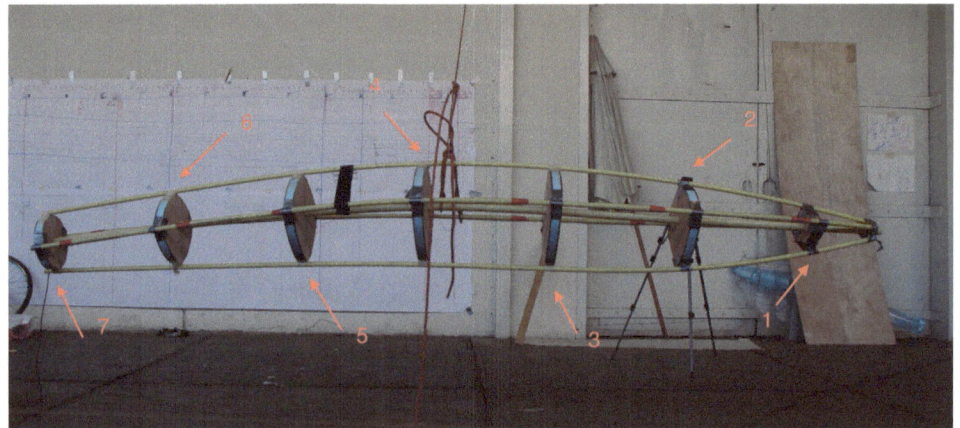

Numbering of the bulkheads

A) Bulkhead Fabrication

Bulkheads are cut from 3.5mm plywood (CAD drawings are available on Protei's git repository) and glued to foam with epoxy. Holes are drilled through the plywood-foam composite for the spines to fit through. Finally, the keel attachment is placed at the bottom of every bulkhead (refer to diagram below).

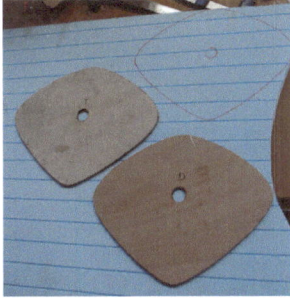
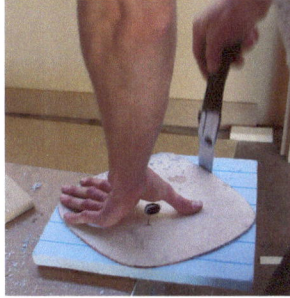

Cutting the bulkheads

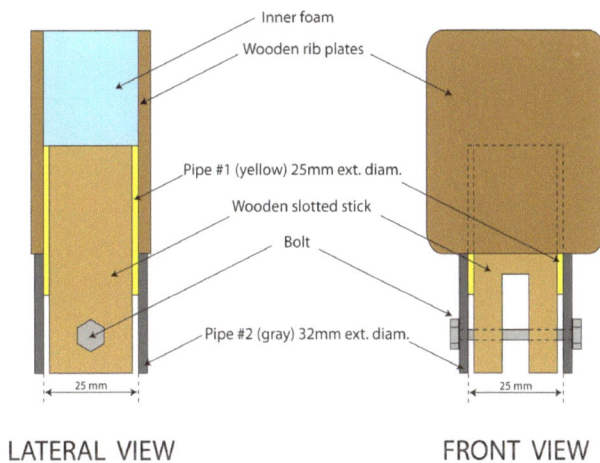

Inner foam

Wooden rib plates

Pipe #1 (yellow) 25mm ext. diam.

Wooden slotted stick

Bolt

Pipe #2 (gray) 32mm ext. diam.

25 mm

25 mm

LATERAL VIEW

FRONT VIEW

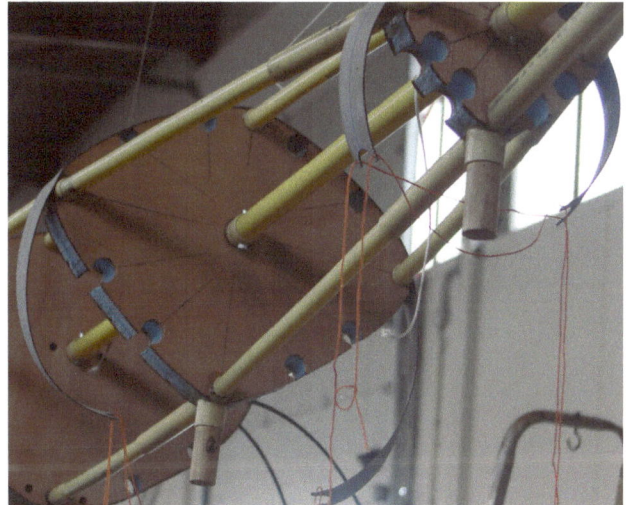

Assembling the keel attachment for each bulkhead

Next, the connection between bulkhead 3 and 4 is built to construct the 3-4 connection pieces from a wood and foam sandwich. In addition to the main sandwich piece, there are 4 aluminum pins and 4 wooden supports that maintain the connection attached to bulkheads 3 and 4. After the parts are fabricated, the wooden supports are screwed into the bulkheads 3 and 4, and the sandwich is connected to these supports using four pins (these pins can be PVC but preferably should be aluminum).

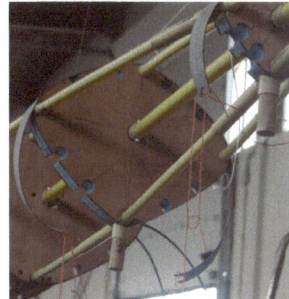

B) Pipe Cutting and Assembly of Spines

The central spine is made up of a 3m long, 22mm extension diameter PVC tube that runs through the body. To constrain ribs laterally, 50mm pieces of PVC (26mm ext diameter) pipe are glued to the bulkheads (see detail below). The spine is passed through the bulkheads and a screw and nut are used to connect the long spine to the 50mm pieces.

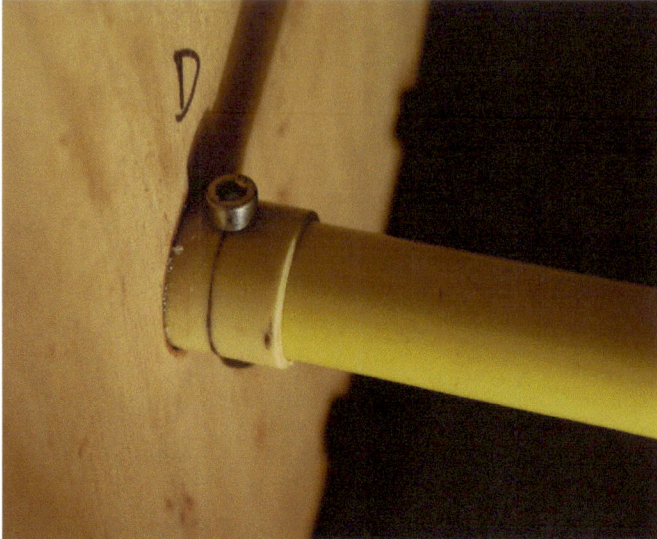

The next step is to cut the PVC tubes for the lateral spines. From now on, this section refers to two different types of tubes: the tubes referred to as "active" *do* have Bowden cables inside (there are 4 of them - one for the top,bottom starboard and port); whereas "passive" spines refer to those that do not have Bowden cables (there are 8 of these).

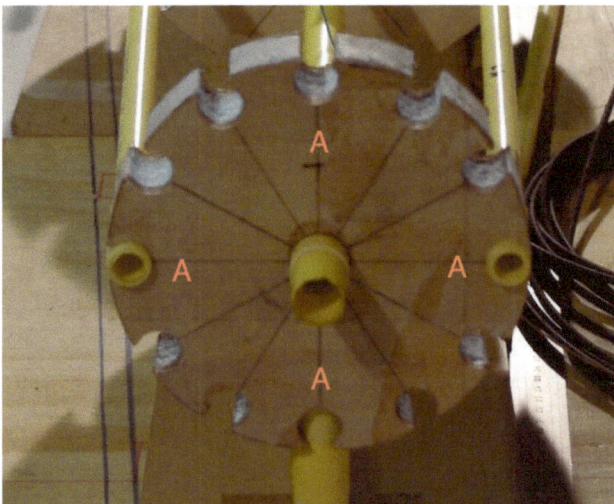

Bulkhead 7 with the slots for the active spines labeled
(the unlabeled ones are passive)

Assembling the Active Spines

Assembling the 4 active spines is a rather complicated task.

The active spines are comprised of two different diameters of PVC tubing: 16mm tubes that slide into a 19mm PVC tube that is glued to the bulkheads using epoxy. See the picture below for detail:

After the spines are in place, the Bowden cables can be routed. The starboard and port Bowden cables are routed first. These cables are routed from the bow and stern to the midships compartment (between bulkhead 3 and 4). The cables are constrained at the bow using another piece of PVC glued into the inside of the spine and a Bowden cable screw. A detail of how the cables are constrained at the bow is shown below left.

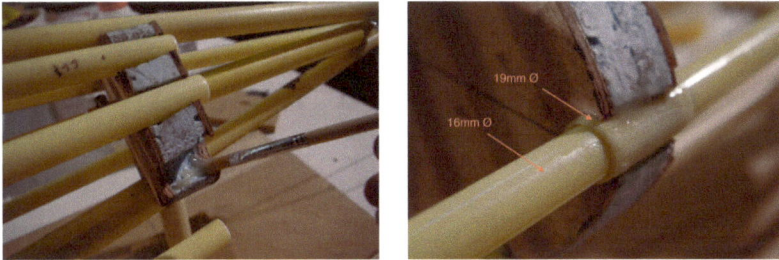

Gluing actuation tubes to bulkheads.

The lateral Bowden cables travel through the spines until the midships compartment where they are routed to the starboard side so they can be then pulled on by the linear actuators. Detail is shown below right.

Constraining of the Bowden cables at the bow.
Note the Bowden cable screws and brown PVC insert.
The stern cable attachments are done the same way.

Detail of the midships starboard compartment showing lateral Bowden cable connections.

Next, the top and bottom Bowden cables are routed. These cables are not meant for actuation but for changing the amount of vertical curvature and to provide some flexibility of the hull in waves. These cables are attached to springs close to the midships compartment.

Detail of the top Bowden cable attachment to the midships section.

Assembling the Passive Spines

The 8 passive spines are assembled in a similar, but more straightforward way compared to the active spines. Pieces of 12mm PVC tube run between each bulkhead and are glued to a 14mm outer PVC tube that then gets riveted to the bulkhead.

Hull alongside the glued passive spines ready to be inserted.

C) Riveting the Hull

After the tubes are put through the bulkheads, pieces of rubber textile are wound around the bulkheads and tied at the bottom. Then rivets are put through every other bulkhead in the passive spine slots. After the rivets are put in, the bulkheads can be dismantled along with their corresponding spines.

Rubber and textile belts around bulkheads with orange tension lines (left) and rivets (right)

D) Adding Flotation to the Hull

There were two ways of adding flotation that were explored. First, airbags were placed inside the body and pressurised so that they fit snugly inside each compartment.

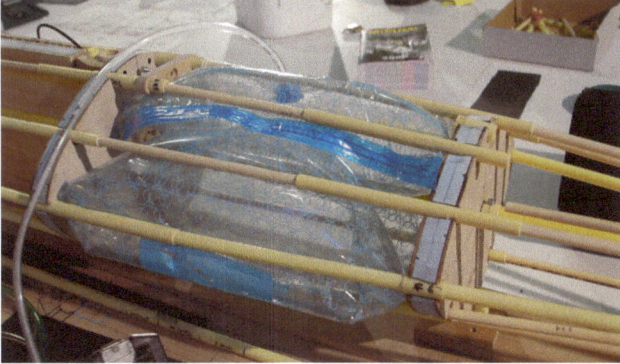

Airbags inside hull

However, the airbags were abandoned because it added too much stiffness to the bending, and the seals on the bags were not reliable as they constantly leaked. As an alternative, Styrofoam slices were used to provide flotation. Circular slices were cut for each section and then adjusted to allow space for boxes, connectors, spines etc. inside the hull.

The Styrofoam slices had the advantage that they provided a fixed amount of buoyancy and did not oppose the bending of the hull.

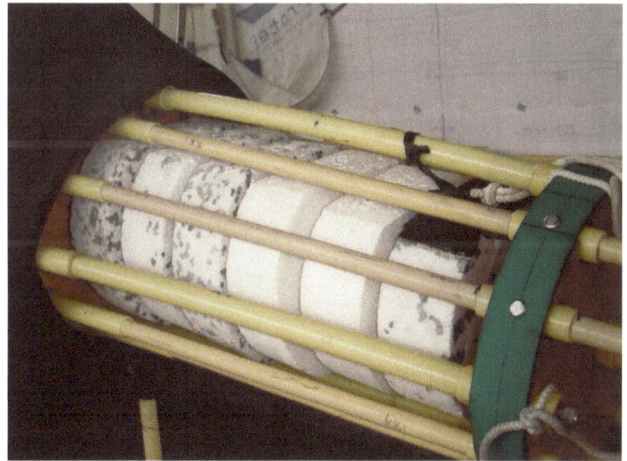

Adding flotation to the hull

Keel

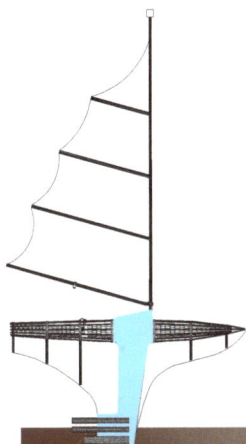

The keel of Protei_006 consists of a rigid main wooden keel roughly midships with the bow and stern sections of EPDM-Foam that flex when the hull bends. The keel extends into the water 1.2m. Its long design provides a long lever arm to right the boat when heeled. Most heavy electromechanical parts like the batteries and linear actuators are placed in the bottom of the keel, which reduces the amount of ballast that needs to be added, hence helping to keep the boat lightweight.

Top: wooden keel
Bottom: flexible keel

Rubber reinforcement on keel's edge

The main rigid keel is comprised of two inner plates of wood sandwiched together by two outer plates. It houses most of the boat's ballast at its lowest point, which keeps the boat upright. Three horizontal slots are cut in the very bottom of the rigid keel, one for the battery and ballast tube, and two for the linear actuators. In the inner vertical plates of the keel, there are two narrow slots running vertically from the horizontal slots, to the compartment in the body of the boat that houses the electronics and the winch. Through these slots runs tubing which holds the wiring from the batteries and the linear actuators in the keel to the control box.

The two outer plates are screwed in at multiple points, this makes the rigid keel simple to disassemble. The flexible keel is fish inspired: more rigid forward and more flexible aft. The forward section is kep rigid with vertical tubes running from the bulkheads whereas the aft section does not have these tubes. Both flexible sections are bendable in 3 dimensions to allow for the hull to move and the vertical tubing prevent the keel from wrapping itself around the hull.

Keel Construction and Assembly

The rigid keel is cut from four wooden plates, two outer plates and two inner plates. The inner plates are the main structural support and also have slots fot the cable routing, and the outer sections sandwhich the keel together. All the plates in the keel are cut to the same template (depicted below and DXF available online).

Cut out for the cable routing

Keel with actuation tubes

2D Drawing of the Keel Cutouts

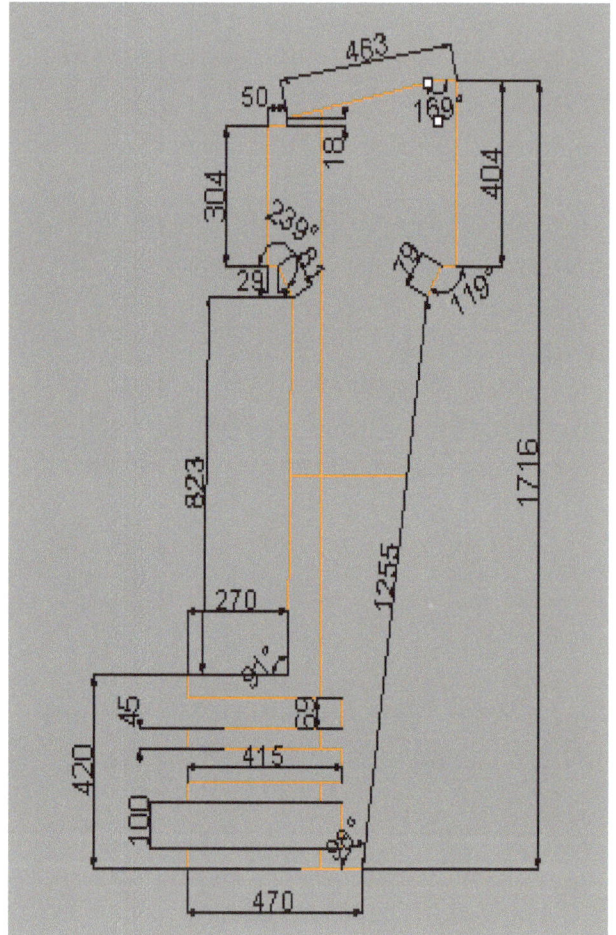

Inside slice of the keel

3 of the 4 slices assembled

The flexible keel is composed of EPDM-Foam reinforced by rubber along its edges (obtained from bicycle inner tubes). The flexible keel is connected to the rigid keel via pins. Vertical tubes that run in the keel are connected to the wooden sticks of the bulkheads, this gives the keel rigidity. Also a small PVC pipe is passed inside the rubber of the leading and trailing edges of the keel, in order to keep the edges in place.

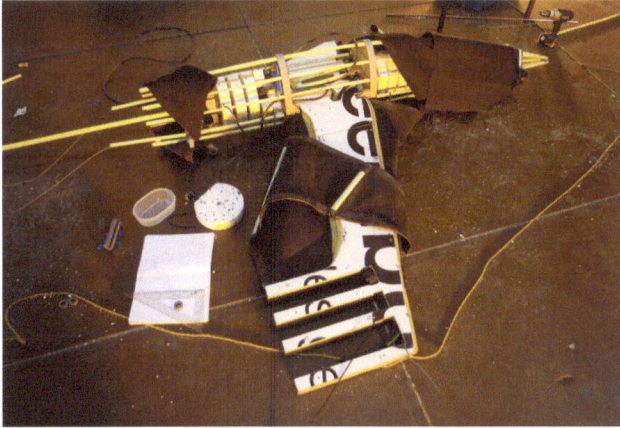

Flexible keel attched to wood keel

Vertical tubes that hold flexible keel to bulkheads

Rigid/flexible skin connection

Skin

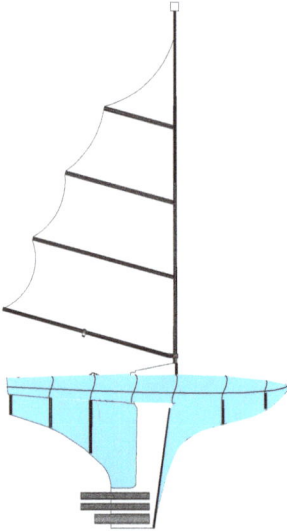

The skin is constructed from neoprene-like segments, organized in an "armadillo armor plate" fashion. Each segment slides over the front of the section it is placed behind, allowing the articulated hull to curve. At the front ends of each segment, the skin is riveted onto the frame (the ribs) of the hull. At the rear end of each segment, it is held in position by slightly pre-tensioned rubber straps. The rubber straps ensure that the entire skin will return to the straight position after the hull is bent. The skin is reinforced on all the edges with rubber to prevent ripping. The skin does not waterproof the hull, but rather it streamlines the hull and adds buoyancy to the vessel.

System overview

The skin is composed of EPDM-Foam (ethylene propylene diene monomer), a neoprene-like material, and consists of six segments draped over the hull and attached to the keel and the frame.

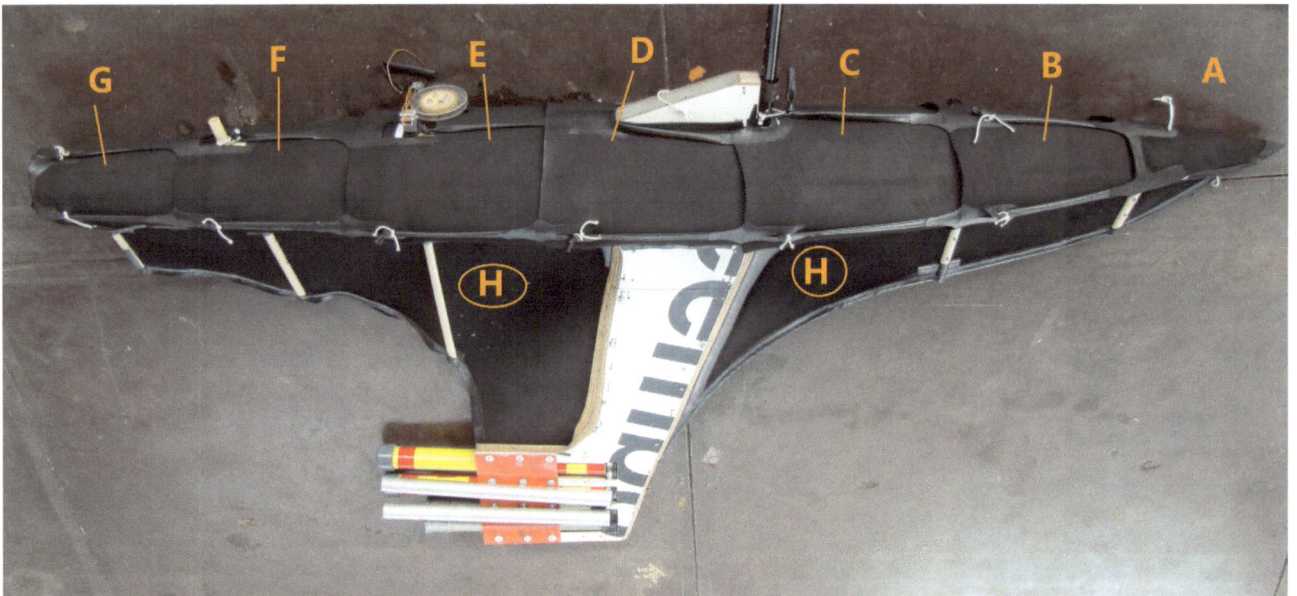

The skin sections of Protei_006

These skin segments are designed to hold the hull tightly, while the elastic feature of the skin material enables the hull to bend in both directions. Skin parts A to G are used to cover the hull and part H is used for steering purpose. At the bottom of the hull, the skin is attached to part H with bolts, nuts and washers. The shape of part H still must be tuned and optimized in order to give a good steering performance.

Patterns to cut these skin pieces are shown on the next page.

Construction patterns

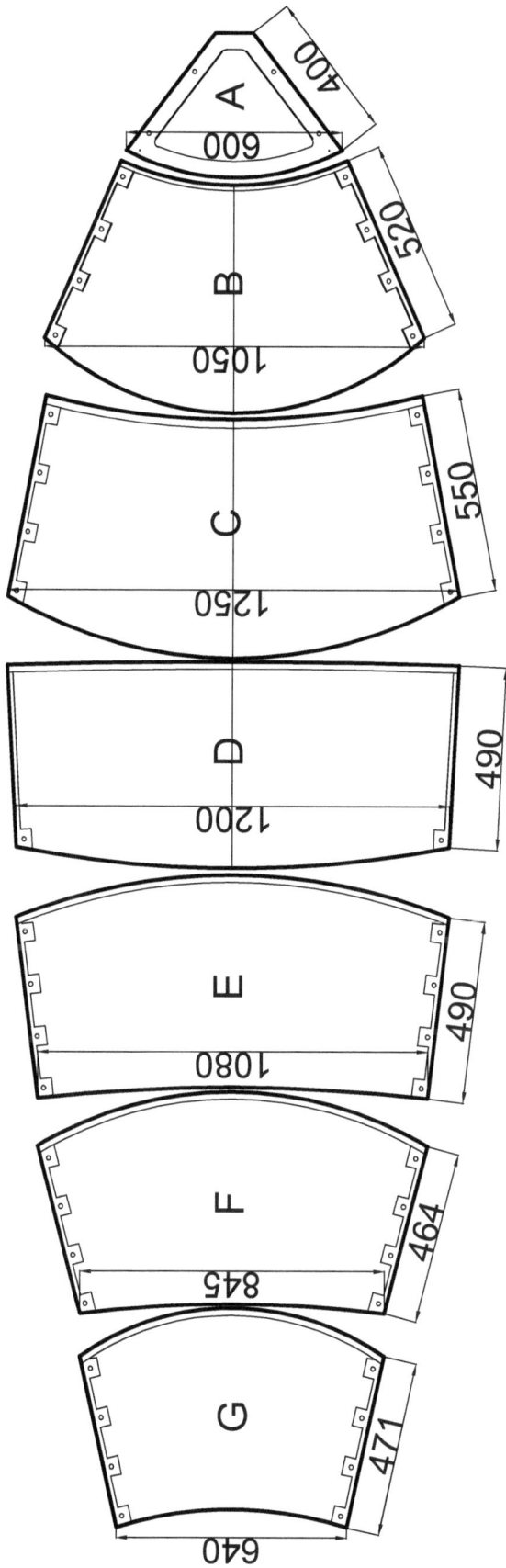

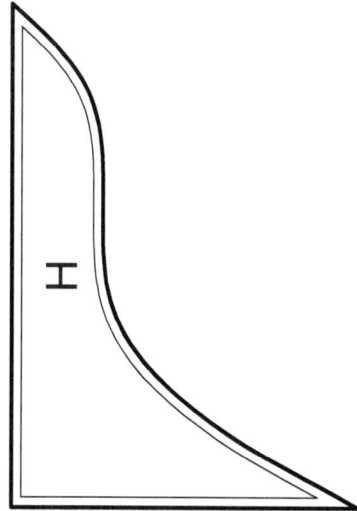

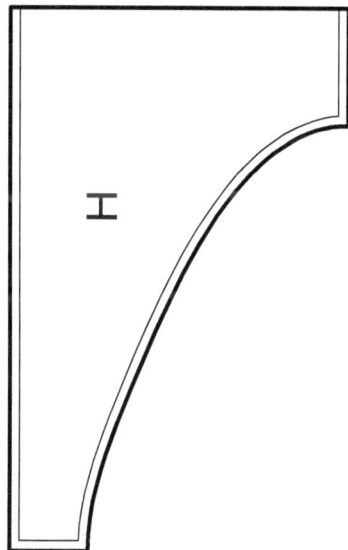

A

400

600

B

520

1050

C

550

1250

D

490

1200

E

490

1080

F

464

845

G

471

640

H

H

Constructing the skin

Riveting

Six straps are cut from rip-stop material, and the lengths correspond to the circumference of each bulkhead (the cross sectional pieces of the hull). We used left-overs from the sail material to create these straps. Each strap is positioned around a bulkhead with a hole drilled through for the rivet. Each rivet is fastened, starting at the bottom of one side.

If the rivet is hard to insert, it can be gently hammered. The strap is pulled tight and worked around the segment. This procedure is repeated for each segment, except for the first bulkhead, on the bow. This one is not riveted. Instead, a few inches of skin is sewn closed around it, to seal the nose so the skin can be slid over the frame. Note that the central segments of skin on each bulkhead should only be attached on the one side of the hull, allowing the skin on the other side to remain loose so the electronics inside the hull can be reached.

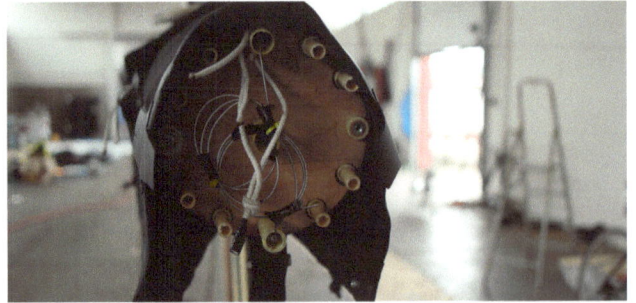

Reinforcing the edges

Because the edges tear rather easily without reinforcement, the skin is reinforced with rubber on every edge section, particularly at the bottom where it connects to the keel, and on the slits where winch and the mast poke through. (These slits are cut afterwards and not shown in the pattern cut-outs).

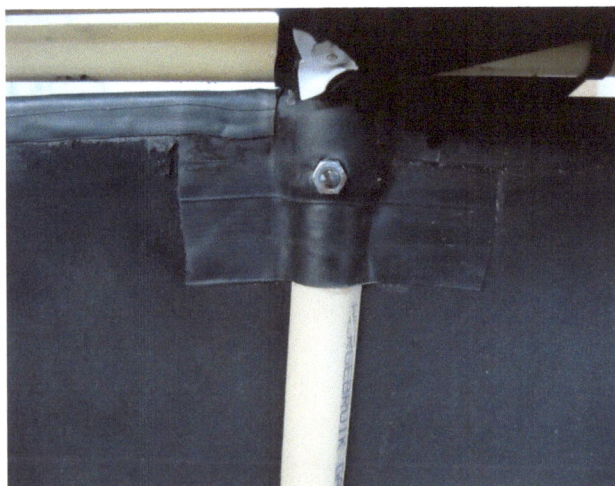

Above: the skin should be gently draped over the hull prior to riveting it to the bulkheads.

Left: the completed rivets. The white strap is the «rip-stop» material.

Above, right, and right-above: bicycle inner tubes have been cemented to the edges of the EPDM-Foam to prevent tearing.

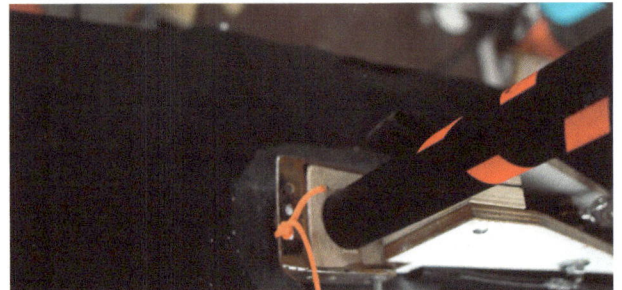

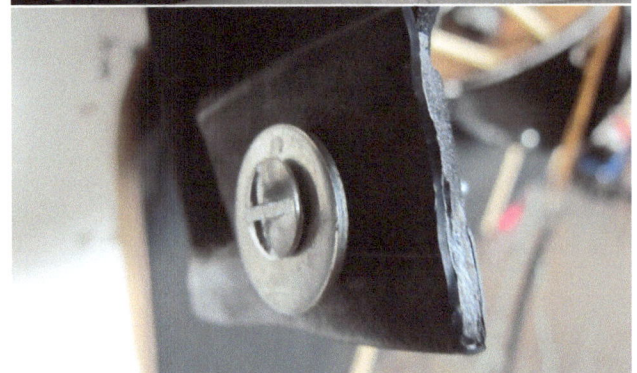

Inner tubes from bicycle wheels are very suitable for rubber reinforcements. The inner tube has a distinct crease, which can be used to slip around the edges where two sections of the skin need attachments, for example, where the washers and bolts go through the keel.

For this section on the bottom of the hull, where the washers penetrate the rubber part of the keel, the inner tubes are cut open as shown in the figure, so that they can be folded around the edge sections of the skin and reinforce the area where the bolts and washers will be attached later on.

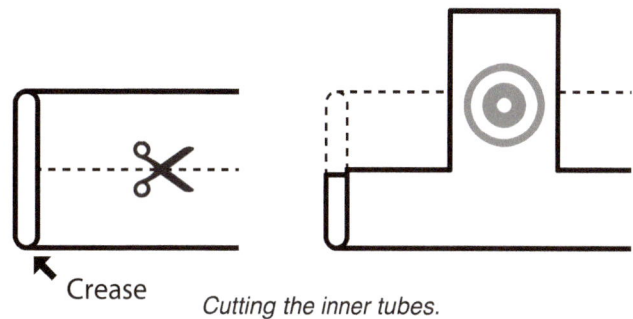

Crease

Cutting the inner tubes.

Gluing rubber and EPDM-Foam

Two-component glue and binder (Rudol KD 4431) are used to glue the rubber to the EPDM-Foam. A few drops of the binder is stirred into the glue until they are well-mixed and a homogeneous color. The rubber is sanded thoroughly and cleaned with acetone. A layer of the glue is applied to both surfaces. After drying for a few minutes, the rubber and the skin are pressed firmly together at every contact point. Applying force is more important than the duration of the pressure. To prevent buckling of the rubber, it is stretched slightly when attached to the EPDM-Foam.

Adding flexible connections

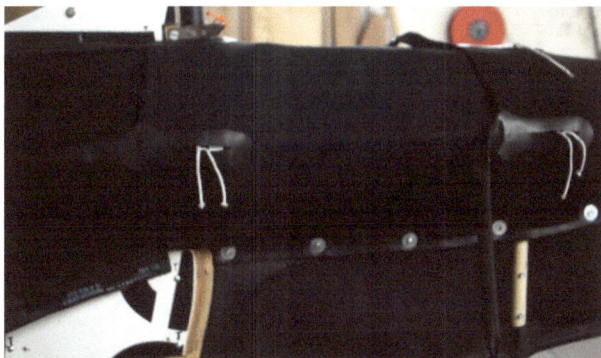

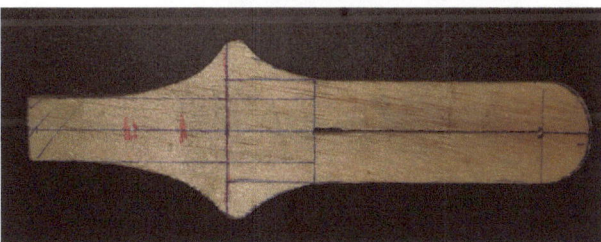

Above: the skin with a connection patch, and the pattern for constructing the patches. Below: slits in the connection patches, and attaching the straps.

Flexible rubber straps are glued to the edges of each skin-segment. Inner bicycle tubes are also appropriate here. On the inner tube, slits are cut by slicing through the inside and outside of the rubber. The edges can be rounded for better gluing. The cut inner tube can be slid over the side of the skin segments and glued to the skin.

The straps hold the skin in position, therefore they must apply a certain tension to the skin. Inner tubes of higher elasticity work best, because if the tension of the skin is too high, the motors will stall when curving the hull.

For extra reinforcement, rubber sheet connection patches can be glued on top of the straps. The stencil (see figure) refers to the shape that the patch should be cut. Once cut to shape, a slit is cut in the middle of the inner tubes and the skin so that the ropes can be pulled through and tied to the rubber strap of each segment in front.

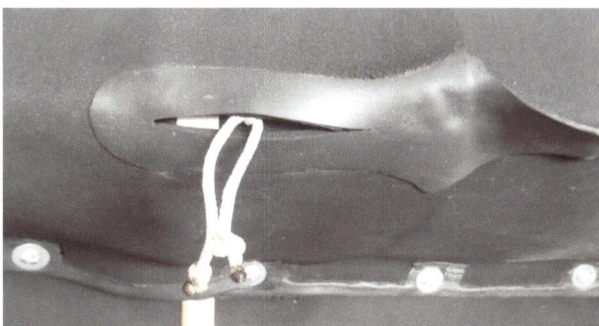

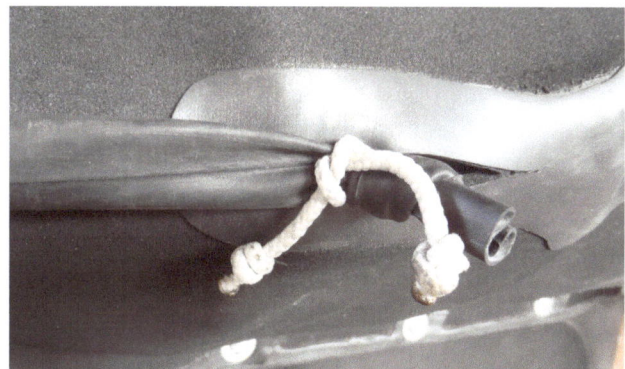

Connection to the keel

Finally, the skin is closed around the hull by connecting it to the keel. The holes are punched through the upper part of the keel and at both sides of each skin segment, through which a bolt is placed, with washers on either sides.

Bolts, nuts, and washers.

Finishing the skin, by connecting through the keel

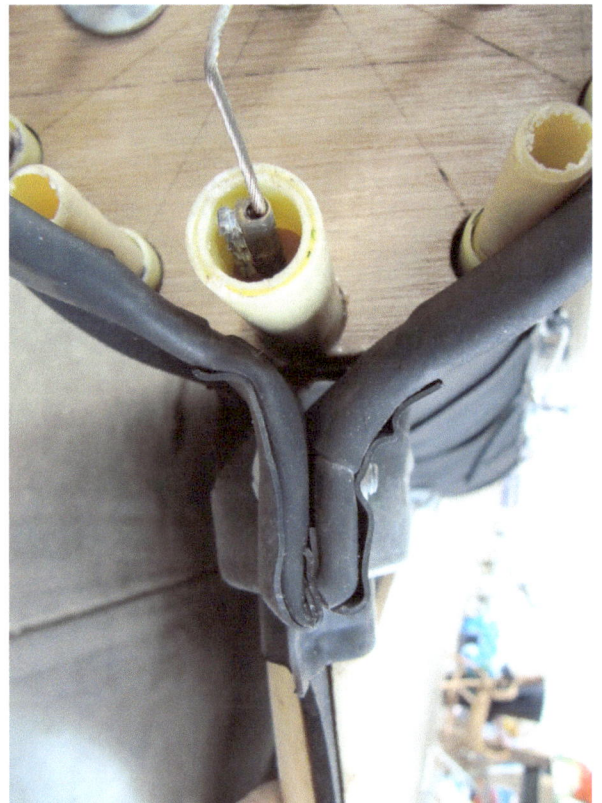
The bolt connection for the skin

Sail

ø3

26.23

96.85

66.65

006

74.27

63.67

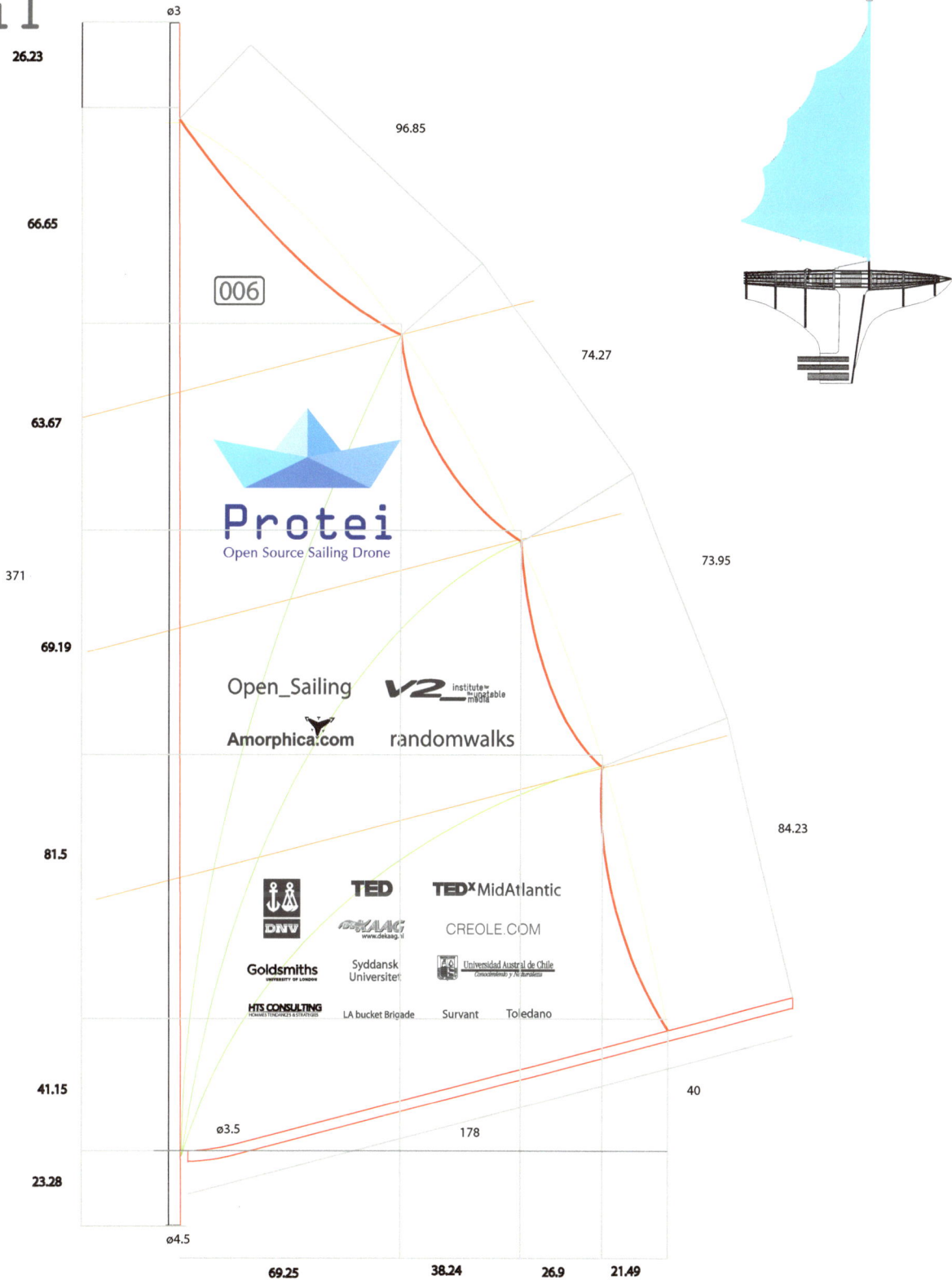

Protei
Open Source Sailing Drone

371

73.95

69.19

Open_Sailing

V2_ institute for the unstable media

Amorphica.com   randomwalks

81.5

84.23

⚓ DNV

TED

TEDˣ MidAtlantic

KAAG
www.dekaag.nl

CREOLE.COM

Goldsmiths
UNIVERSITY OF LONDON

Syddansk
Universitet

Universidad Austral de Chile
Conocimiento y Naturaleza

HTS CONSULTING
HOMMES TENDANCES & STRATEGIES

LA bucket Brigade   Survant   Toledano

41.15

40

ø3.5

178

23.28

ø4.5

69.25   38.24   26.9   21.49

We made two research versions of the sail (see next page). For the third and final sail of Protei_006, we sim-ply recycled a larger sail fabric that was given to us by the Kaag Watersport Academy. The general design is inspired by a bat wing. Thin flattened PVC pipes give structure and support to the sail in velcro-enclosed pockets. The logos are spray-painted using adhesive stencils. Protei Logo by Roman Yablonski.

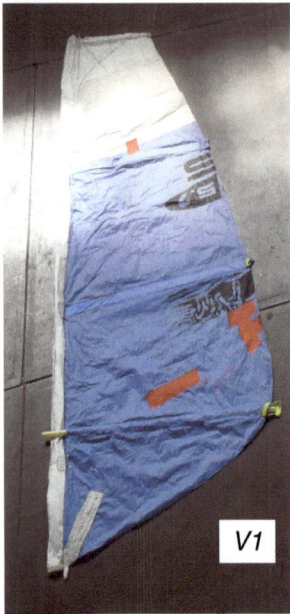

V1

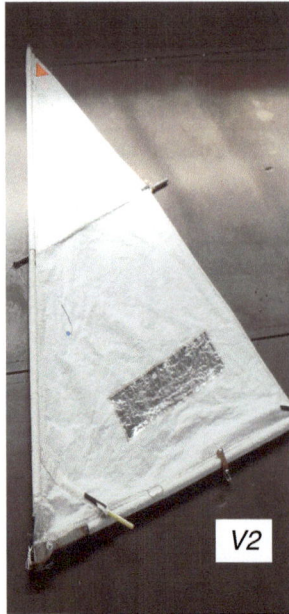

V2

Protei, biomimicry inspirations

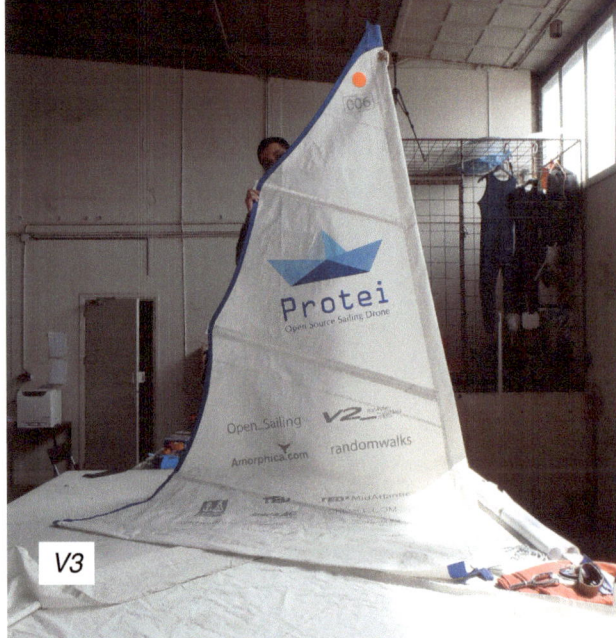

V3

Sail winch

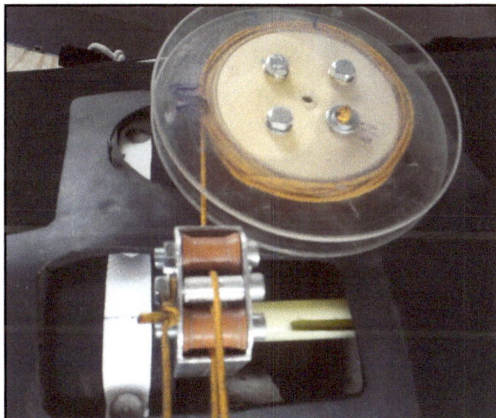

Drum and main block wired

3D view of the winch

Sail winch overview

The sail winch provides the control of the main sail of the boat, by reeling in or out the main sheet. It consists of a drum, which shaft is coupled to a DC motor, a lid for water-proofing, and three blocks for wiring.

The main block is fixed to the hull and consists of three rolls that distribute the main sheet to the second block, a traditional double pulley.
The last block is a simple shackle connected to the boom by a wooden clamp.

Between the double pulley and the shackle, there is only one sheet that is doubled by a shorter rubber band, in order to keep tension at any time.

Sail winch assembly

The assembly for the winch is divided into the following steps :

1. Assembling and fixing the winch motor box
2. Assembling and fixing the main block
3. Assembling the boom clamp
4. Wiring the drum, the main block and the double pulley
5. Wiring the shackle

Assembling and fixing the winch motor box

The winch motor box is composed of :
- a 12V DC geared motor. We used a car windscreen wiper motor (see picture).
- a lid and the associated box for waterproofing.
- a wooden support, that will be screwed to one of the boat's ribs. This support has a hole for the shaft, three holes for the vertical screws, and a hole for the power cables.
- a small piece of PVC tube glued inside the hole for power cables to let the power cables go through the lid.
- an aluminum coupling that fits inside the hollow shaft and around the motor shaft.
- an aluminum hollow shaft with a coupling for the motor. Ø28mm, length : 125mm.
- a wooden drum support (Ø58mm, 2cm long) with a centered hole of Ø28mm.
- a drum : two round slices of Plexiglas (Ø14cm) and an inner drum of Ø8cm.

We used three vertical screws to hold the motor, the lid and the wooden support together. In order to provide efficient coupling with the motor, we drilled a Ø4mm hole trough the motor output shaft and we threaded it (M4 standard metric — ISO 965). We used two M4 screws to hold the motor shaft, the coupling and the aluminum shaft.

The wooden drum support is then screwed horizontally to the top of the aluminum shaft. The drum is then screwed vertically to the support. In order to avoid tearing the skin or increase jamming probability, the screw heads are counter-sunk.

12V windscreen wiper motor

Motor, shaft, coupling, lid and wooden support

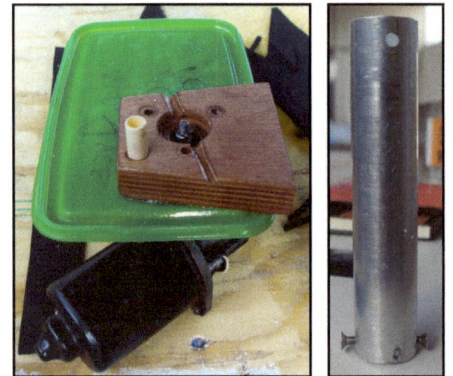
Assembled   Aluminum shaft

Motor, lid, support and aluminum shaft

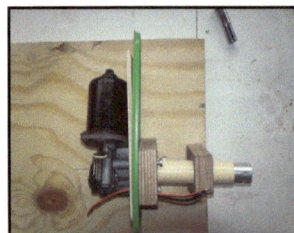
PVC Guide added

Drum support added

Drum added

A flexible rubber tube running to the control box can then be clamped to the output of the PVC tube, to provide waterproofing[1].

---

1   The box we used did not provide sufficient waterproofing. We are working on implementing a more watertight housing.

Assembling and fixing the main block

Here is a 3D view of the main block:
- The central hole (Ø20mm) is designed to let one of the skeleton spine go through.
- The secondary hole (Ø9mm) is designed to access one of the bolts that holds the main block to the rib.
- The back holes (Ø5mm) are also designed to hold the main block to the rib.
- The lateral hole (Ø3mm) is the entry point for the sheet coming from the drum.

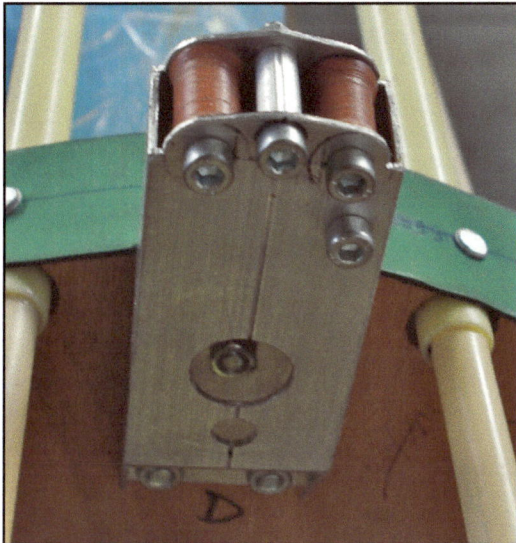

3D isometric view of the main block

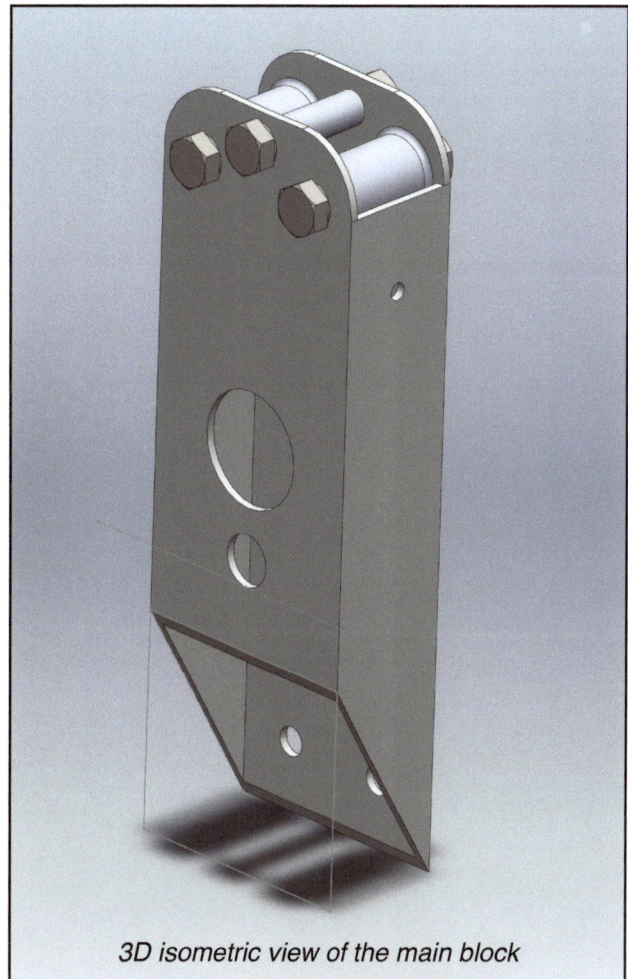

Photo of the main block, before wiring

Assembling the boom clamp

Here is a 3D view of the boom clamp :
The shackle is attached to the Ø7mm hole.
A bolt is going through the Ø6mm hole to clamp the boom, which goes through the Ø32mm main hole.

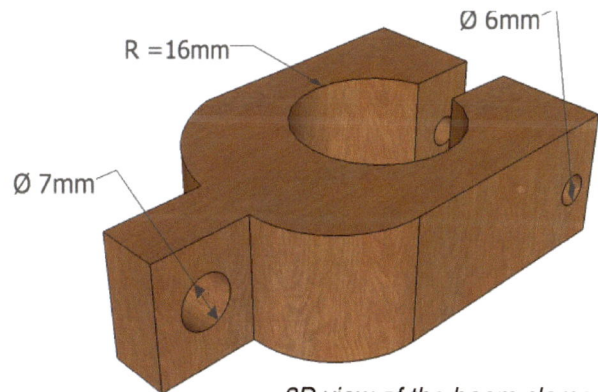

3D view of the boom clamp

Wiring the drum, the main block and the double pulley

Attach the sheet around the drum and follow these steps :
1. Insert sheet into the lateral hole from the main block.
2. Go around one of the main rollers.
3. Go around one of the double pulley rollers.
4. Go around the second main roller of the main block.
5. Go around the other double pulley roller?
6. Attach the end of the sheet to the central roller of the main block.

Wiring the shackle

1. Fix the shackle to the Ø7mm hole of the boom clamp.
2. Wire the shackle to the double pulley with a 10cm long sheet.
3. Cut a 7cm piece of rubber band and attach it between the shackle and the double pulley. Make sure it is shorter than the sheet, but long enough not to break when tension is applied (the sheet should then support the tension)

Linear actuators

The linear actuators provide the tension on the Bowden cables necessary to bend the hull. As this is the primary means of maneuverability on Protei, the linear actuators are quite important. Two identical linear actuators are constructed for Protei_006, one for the bow and the other for the stern. This allows us to control the bending of the stern and hull individually.

Inner structure of the linear actuator

PVC tubes are used as both the structural frame of the actuator and the water proof barrier. The actuators slide into the bottom of the keel, just above the battery tube.

Housing for the linear actuator

Mounting the linear actuator to the keel

Components

1. Limit switch (push button)
2. Moving carriage
3. Aluminum U-channnel guide
4. Threaded shaft
5. Bowden cable
6. Bearing
7. Wooden support part B
8. Soft coupling tube and hose-clamp
9. Bowden cable housing
10. DC motor with gear box and clamp (from electric hand drill)
11. Magnet sensor
12. Bowden cable guiding nut
13. Water cooling tube
14. Wooden support part A
15. Wooden support part C

Operation

Two Bowden cables are fixed complimentarily to the moving carriage. In this way, when the carriage moves one direction, one cable is lengthened and the other is shortened. As the DC motor rotates the threaded shaft, it causes the carriage to move linearly. Pulling on one cable constricts one side of the spine, causing the boat to bend.

The limit switches are used to detect the extreme position of the carriage and prevent the carriage or motors from damaging the linear actuator. When one of the limit switches is triggered, the motor driver is automatically disabled, as described in the electrical section. A magnet attached to the shaft, together with a hall-effect sensor, is used to count the number of rotations, providing us with the position of the carriage.

Details of the flexible joint connection between the planetary gearbox and the threaded rod

Details of the linear actuator, showing motor mounting, the cooling tube, and cable routing

How to build

Wooden support A (part 14)

Wooden support B (part 7)

*Above: top view
Left: side view*

Wooden support C (part 15)

Wooden supports for the linear actuators

All the wooden supports have the same outer diameter so that they can be mounted into the same tube and keep the shaft concentric. One of the disadvantages of these wooden parts is that although they are very easy to manufacture, they are not strong enough to endure strong force. We have already switched to aluminum materials for these supports. The pictures above are included for background about the original linear actuators.

Since the motors are sealed in a small tube, there is no air circulation to cool the motors. The water cooling tubes provide a passive flow of water around the motor while the boat is moving, to conduct heat away from the motor. An inlet funnel helps water flow through the cooling tube.

Future actuators

These linear actuators have several deficiencies that should be mentioned. Wood weakens and degrades in long term use (as can be seen already in the photos above). Furthermore, the support pieces must be attached to the tube by drilling from the outside, creating additional problems of water sealing. This has been sealed over with a very strong waterproof red tape, as can be seen in the photos, but a better solution is necessary. For this reason, we have begun the design of a new linear actuator system.

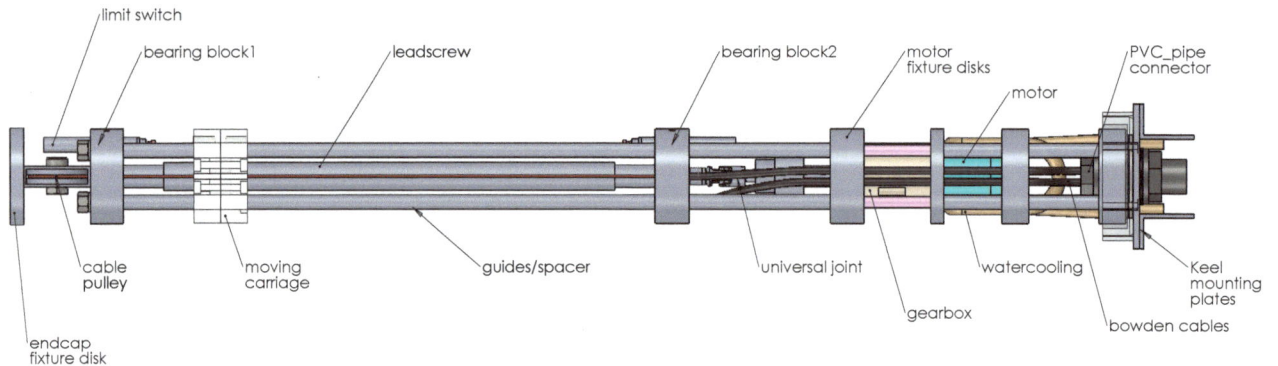

The new linear actuator design uses solid aluminum supports instead of wooden supports, which improve reliability and add weight to the keel. The number of supports has also been increased for a more robust structure. The hose clamp based sealing system has been replaced with an easily removable end cap, which is fixed to the keel, so that hose connections for electrical and mechanical cables to not have to undergo any stress, and will remain waterproof. Horizontal aluminum tubes act as guide rods inside the actuator, and house electrical cables to prevent damage and short circuits.

The carriage jumps off of the threaded rod before reaching the limit of its travel, so that in the case of a limit switch failure, no damage is done to the actuator. The diameter of the water cooling tubes has been increased, and the aluminum disks also act as fins to increase the ability of the actuators to dissipate heat. A pulley at the end of the Bowden cable reduces friction and adds play, preventing bunching or kinking of the cable inside the actuator. The flexible joint between the planetary gearbox and the threaded rod has been replaced with a more robust universal joint.

Additional drawings and dimensions can be found on Protei's git repository at: https://github.com/Protei/Protei-005-6

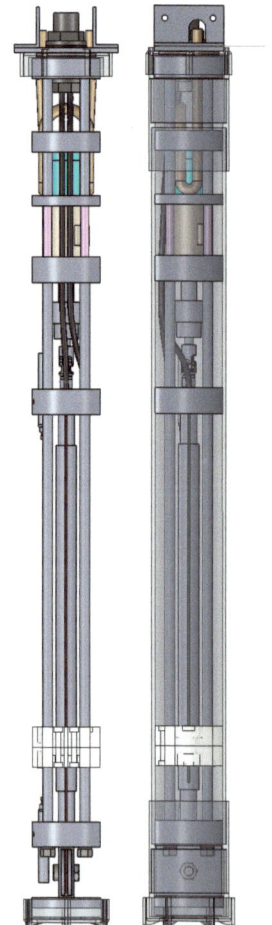

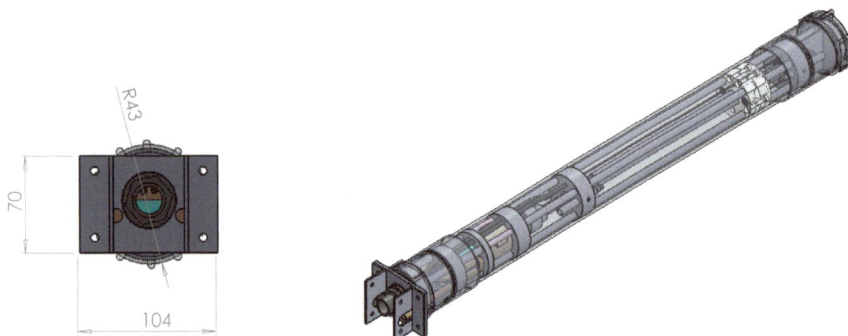

Cable routing

The image below shows the cable routing for the new linear actuator system.

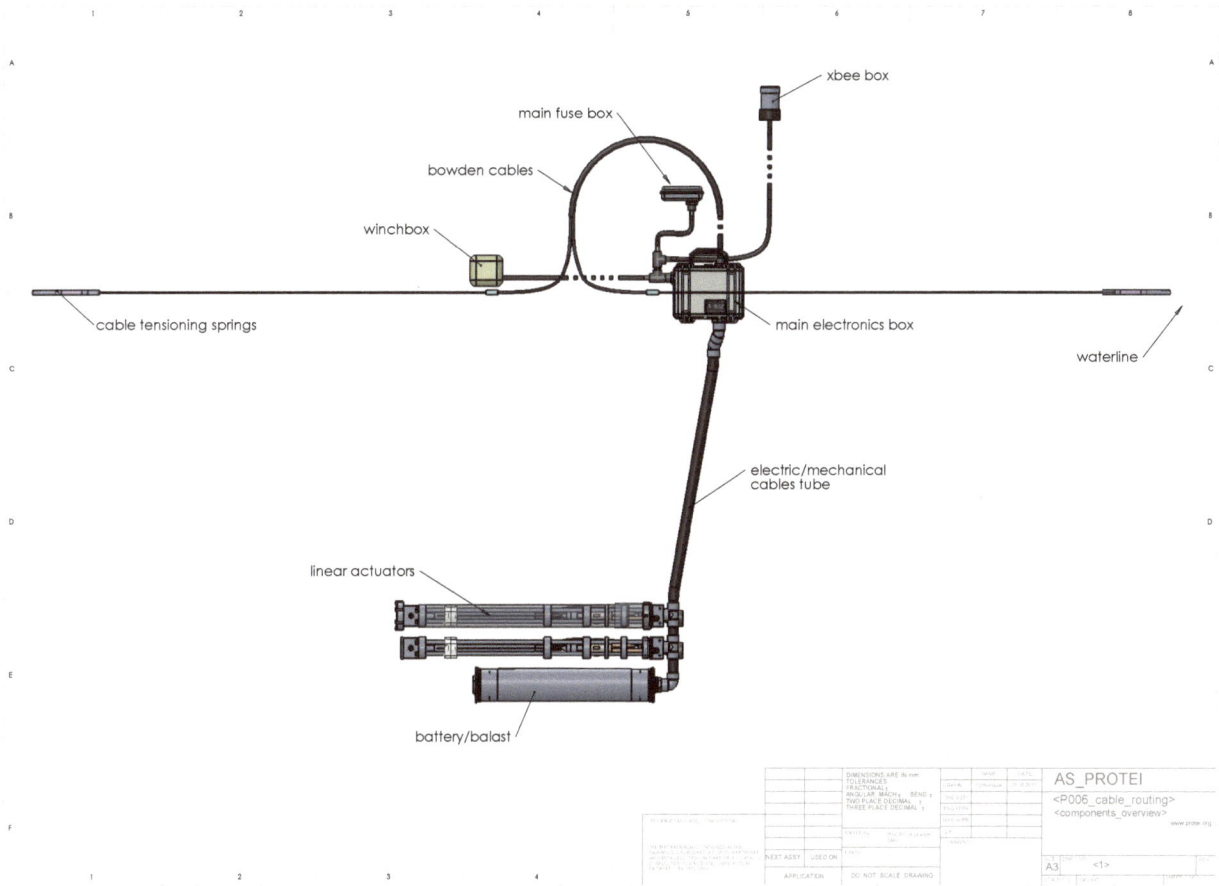

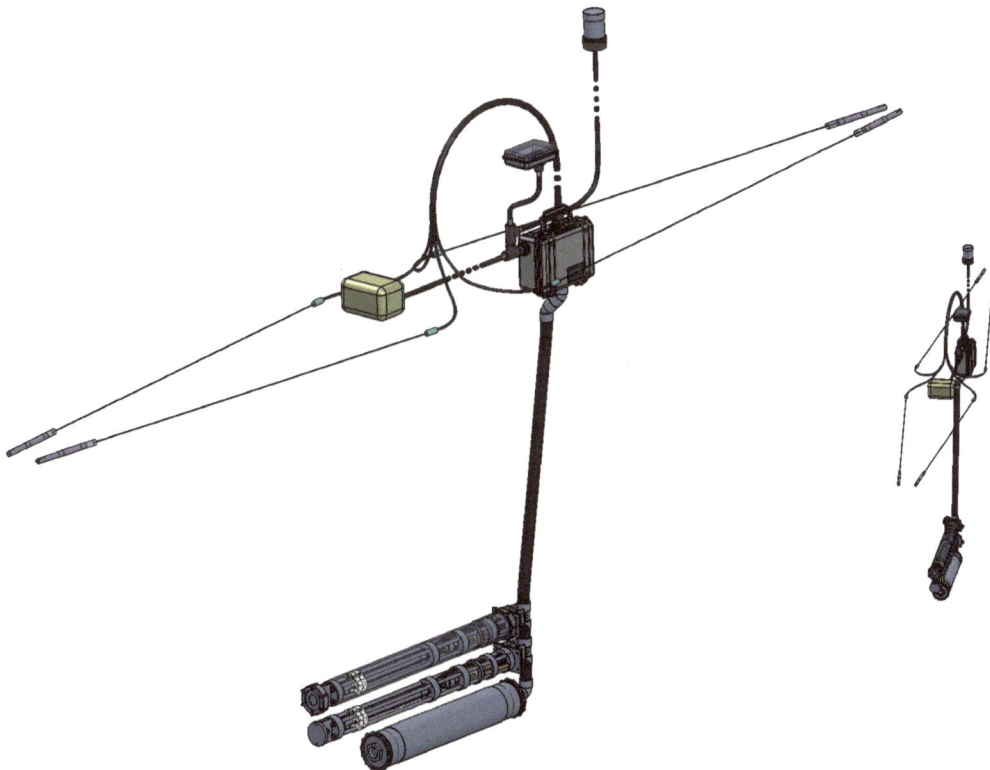

Electrical Design

Protei's electrical system is designed to accomplish two tasks: communicate reliably with an onshore transmitter, and precisely control the on board actuators.

Communication is accomplished using Xbee modems, which allow for medium range (maximum 500–1000m) bidirectional wireless communication.

Two of the actuators control linear movements, these are located in the keel. Through the use of limit switches and a rotational sensor, these can be position-controlled in closed loop feedback. The third actuator, the sail winch, can only be open-loop-controlled at this stage of the development, though connections have been built into the microcontroller to allow for future closed-loop-control using a rotational sensor.

The diagram below shows a schematic representation of the data and power connections between components on Protei_006. It also provides a not-to-scale representation of the physical location of these items. Each item is described in further detail on the following pages. Larger schematic images can be found in the appendices, and source code can be found on Protei's git repository, at: https://github.com/Protei/Protei-005-6.

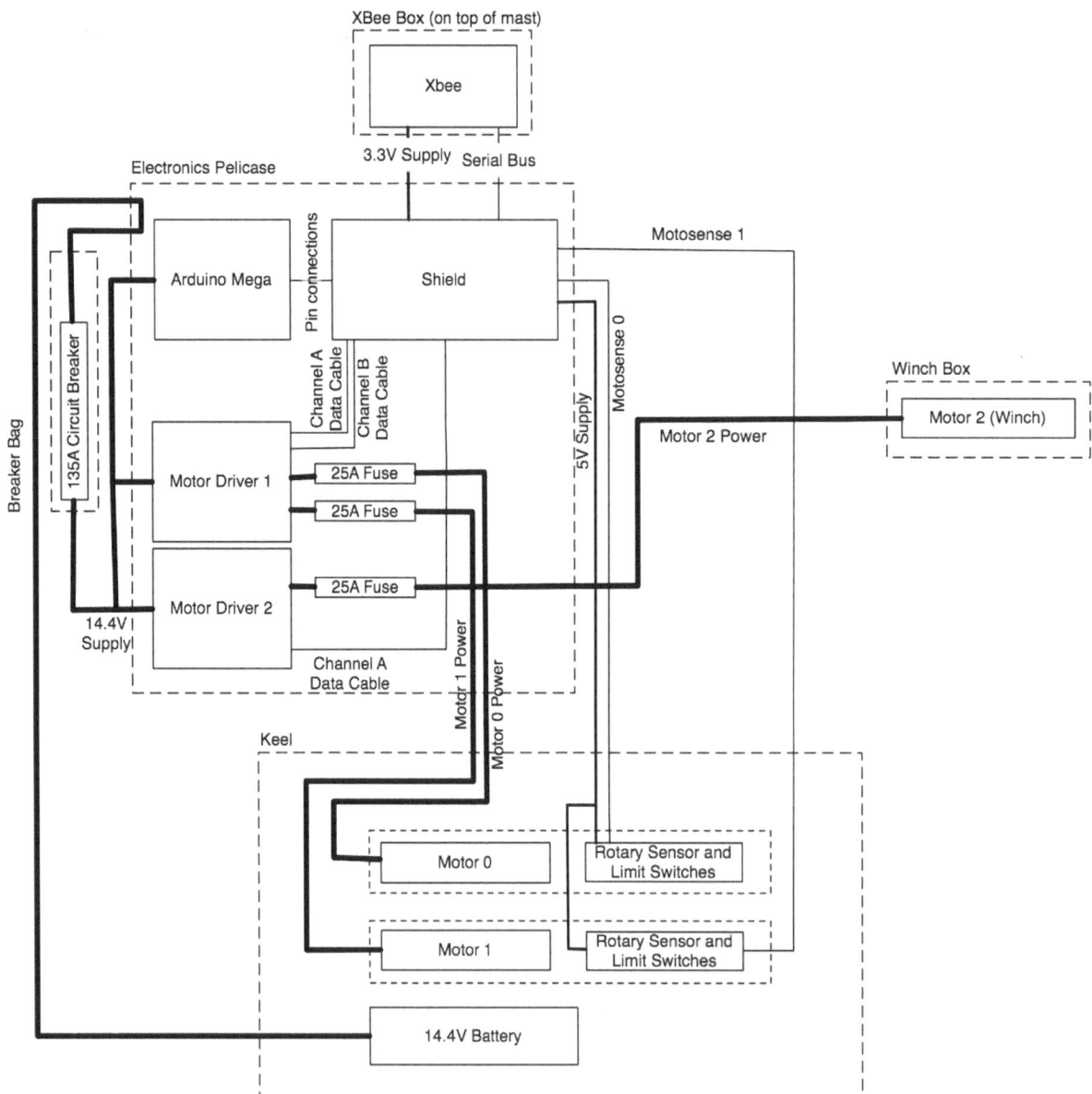

XBee Box (on top of mast)

Xbee

3.3V Supply   Serial Bus

Electronics Pelicase

Pin connections

Arduino Mega

Shield

Motosense 1

Motosense 0

Channel A Data Cable
Channel B Data Cable

Breaker Bag

135A Circuit Breaker

Winch Box

Motor 2 (Winch)

Motor Driver 1

25A Fuse
25A Fuse

25A Fuse

5V Supply

Motor 2 Power

Motor Driver 2

14.4V Supply

Channel A Data Cable

Motor 1 Power

Motor 0 Power

Keel

Motor 0

Rotary Sensor and Limit Switches

Motor 1

Rotary Sensor and Limit Switches

14.4V Battery

Cable routing

Routing cables between the different dry compartments on Protei_006 is actually quite complicated. Each box has one or more short pieces of PVC pipe inserted into it, and glued and waterproofed. A flexible hose is clamped around the rigid pipe, and electrical wires are forced through the flexible hose. In this way, we can create waterproof connections without the expense of pricey waterproof connectors. The diagram below shows a schematic of these wires and the hoses the connect the dry compartments.

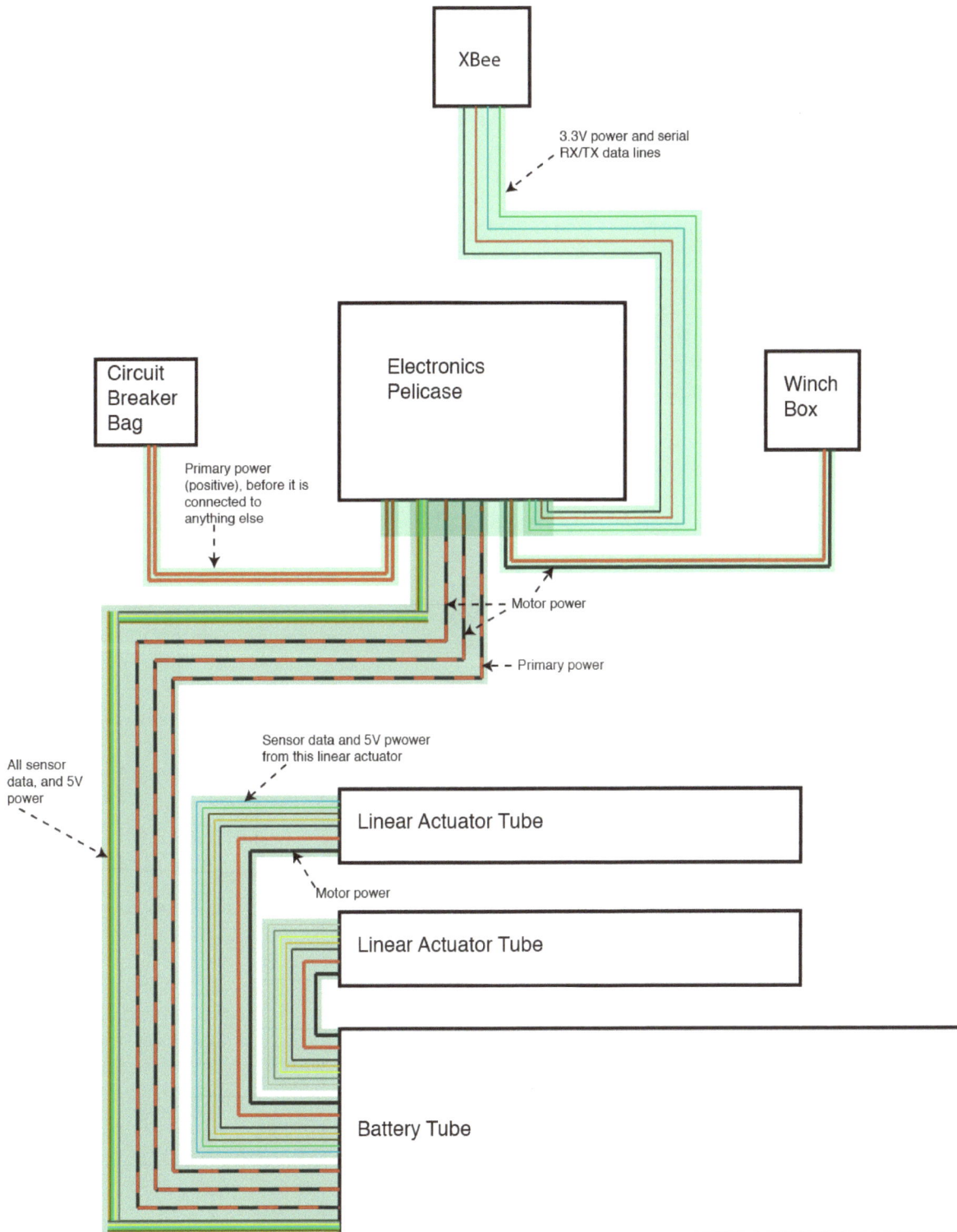

XBee

3.3V power and serial
RX/TX data lines

Circuit
Breaker
Bag

Electronics
Pelicase

Winch
Box

Primary power
(positive), before it is
connected to
anything else

Motor power

Primary power

All sensor
data, and 5V
power

Sensor data and 5V pwower
from this linear actuator

Linear Actuator Tube

Motor power

Linear Actuator Tube

Battery Tube

This cutaway view of the keel shows the channel that the hose runs through, inside the keel.

The wires indicated on the diagram are the actual wires that run inside Protei_006. Thick red and black lines indicate the 14 AWG power cables, and thinner colored lines are small 22 AWG signal wires. The color code is completely arbitrary, the only thing that is important is to be self consistent. The signals in the sensor bundle are: 5V, ground, and two limit switches and a rotational sensor for each actuator, for a total of eight signals. This also happens the number of signal lines in standard Cat5 cable, so we used a Cat5 cable (not in twisted pair configuration) and an RJ45 connector for this connection.

The largest cable hose, running from the Battery tube to the electronics box, runs inside the keel. The hose to the Xbee runs inside the mast, and other cables are simply routed around spines and bulkheads as convenient.

An example of the waterproofing system for connections is shown below:

Above left: a PVC tube connector for an early prototype of the electronics box.

Above: the Xbee tube, showing the hose connection there. Note that the portal is always a rigid tube, with a flexible tube then clamped (not shown) to the rigid tube.

Left: the planned future cable routing for the new linear actuators. See the Linear Actuator section for more details.

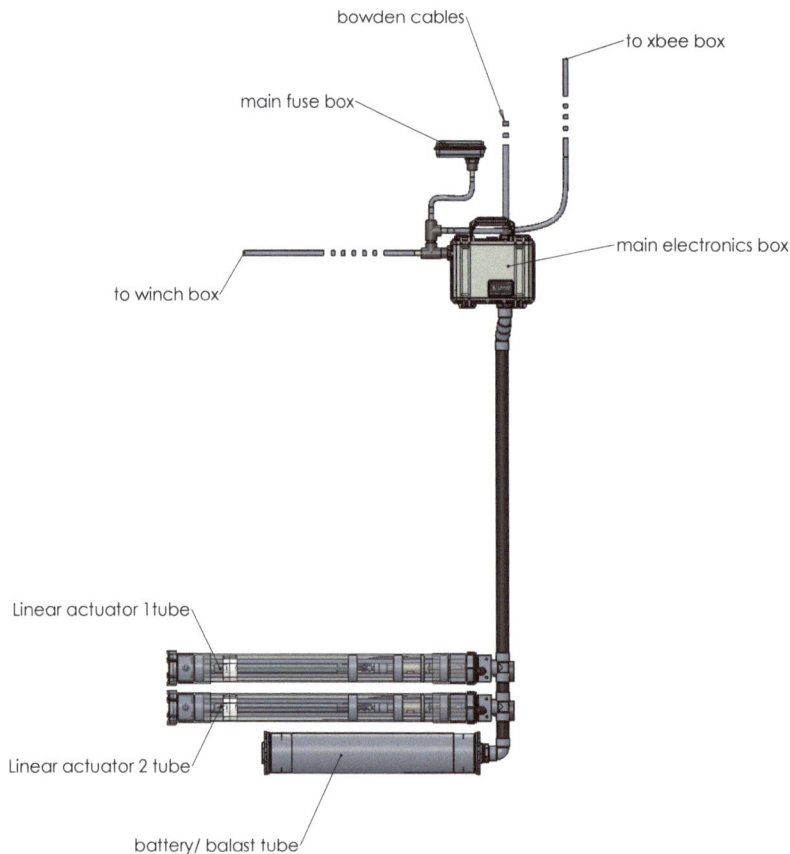

bowden cables

to xbee box

main fuse box

main electronics box

to winch box

Linear actuator 1 tube

Linear actuator 2 tube

battery/ balast tube

Actuation system

The linear actuators on Protei provide the pulling force necessary to bend the boat. The mechanical design has already been described above, this section focuses on the electrical design of the actuators. Electrically, there are two separate mechanisms — the driving mechanism, and the sensing mechanism.

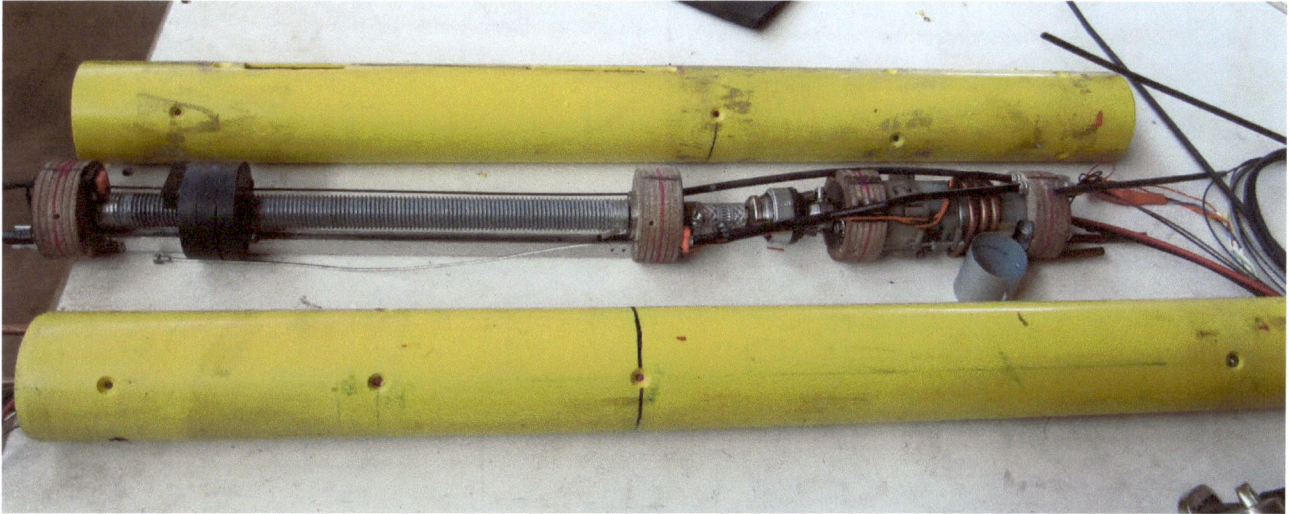

Driving

The motor is driven through an H-Bridge motor driver, purchased from Elechouse. The manufacturer claims that it will work with 12V motors at 50A continuous and 100A peak current. We have not thoroughly tested these claims.

As shown on the right, each channel has six logic inputs — VCC, GND, EN (enable), RPWM ("right pulse width modulation" — this drives the right hand side of the H-Bridge), LPWM ("left pulse width modulation"), and DIS (disable). DIS is always left unconnected, as we do not need the functionality that it provides. VCC is connected to +5V, GND to the GND of the microcontroller, and RPWM and LPWM are connected to two PWM capable output pins from the microcontroller. EN is connected as described below in the sensing section. 12V straight from the main battery is connected to the POWER header on the opposite side. Each motor is connected to the controller in series with a 25A fuse to prevent the motor from damaging the motor driver.

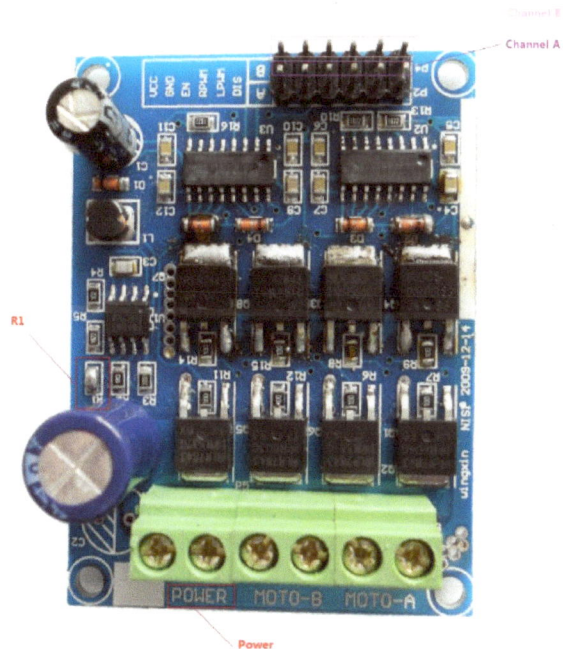

The motor driver board used for driving the actuators on Protei.

Sensing

Sensing is accomplished through two mechanical limit switches, on either end of the actuator's linear range, and a Hall Effect sensor, which detects the rotation of a magnet attached to the motor's shaft. A schematic of the sensing mechanism is shown below. This is, of course, duplicated for each actuator. An Eagle Layout Editor file for this schematic, and all other referenced electrical drawings, is available on the Protei git repository: https://github.com/Protei/.

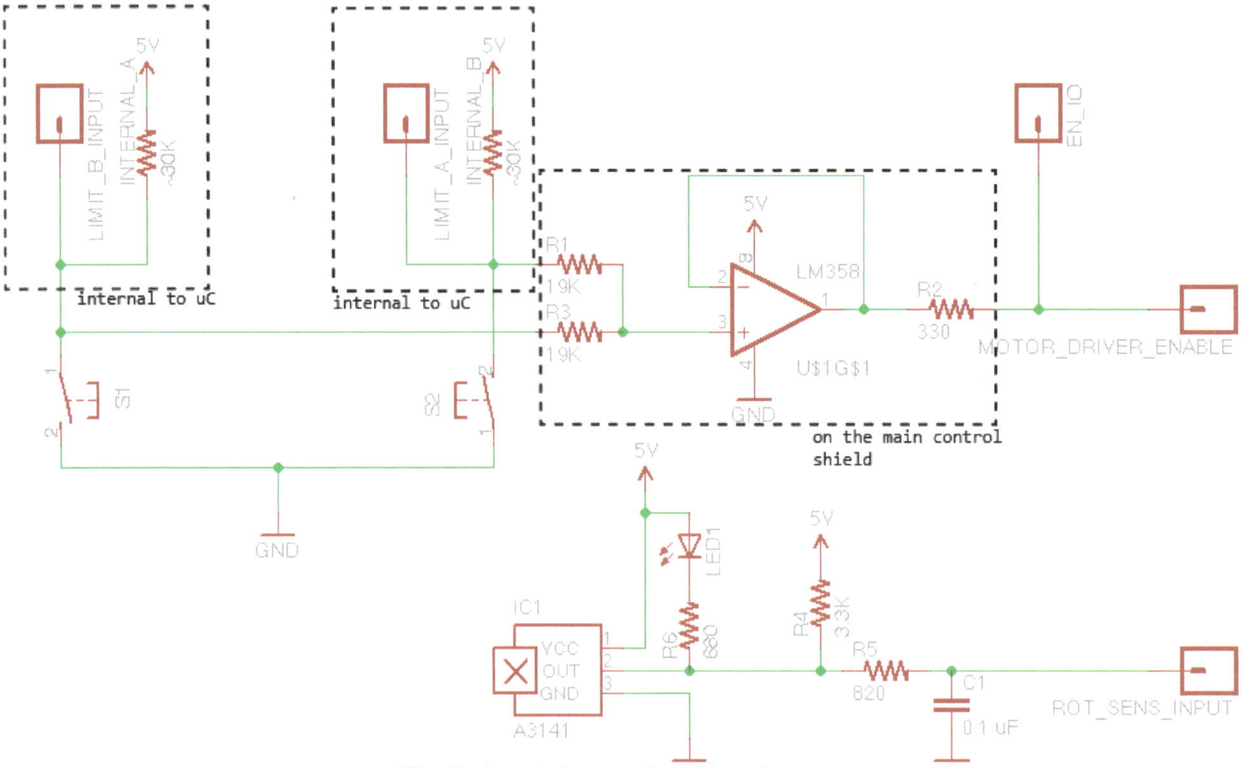

The limit switches and rotational sensor.

The most complicated part of the above schematic is the limit switch function. The limit switches have two functions: the microcontroller can detect the state of each limit switch individually; and, if either switch is depressed, the motor driver will be disabled, until the microcontroller re-enables it. This is implemented through use of the tri-state logic capability of the microcontroller digital I/O. When the EN_IO pin is set to high impedance (input) mode, the value of the motor driver enable pin is entirely set by the state of the switches. When one of the switches is toggled, the enable pin goes low, and can be reasserted high by changing EN_IO to output mode. In this way, the actuator is protected against software failure, and does not solely rely on the microcontroller for detecting switch presses. Obviously this only works if the default state of EN_IO is high impedance.

An image of the limit switch used in the old design of the linear actuators.

The rotary sensor uses an Allegro A3141 hall effect sensor to detect the presence of a magnet on the motor shaft. R4 serves as a pull-up resistor and R5/C1 is a lowpass filter. The output is briefly set low once per revolution, as the magnet passes the sensor. This can be read with a digital input on the microcontroller.

Communication system

The communication system uses two Xbee Pro 900 Series 1 devices for short range (~1000m) bidirectional radio communication. These communicate using the following simple protocol. The source code for the transmitter is in ArduinoRC.pde, and for the receiver, in `ArduinoControl/comm.h`, in the Protei git repository.

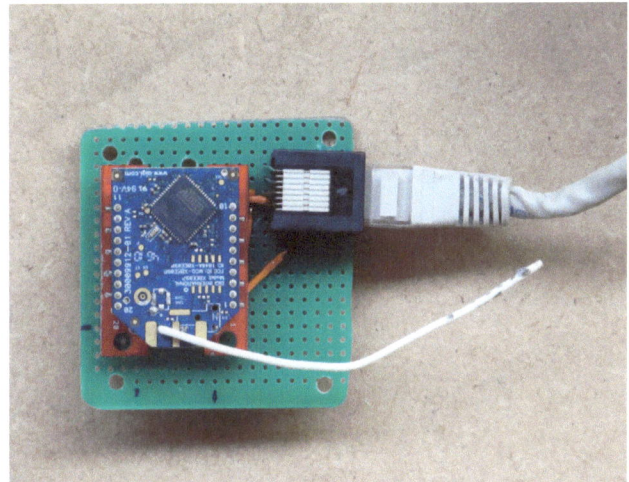

Every 100 ms, the joystick controller (transmitter) sends three pieces of information to Protei: the position of its left joystick, the position of its right joystick, and the status of its buttons (127 if no buttons or all buttons are pressed, 0 if just the left button, 255 if just the right button). These three bytes are split into two half bytes of 4 bits each, and each half byte is encoded with a Hamming7,4 code plus an overall parity bit to improve reliability of communication.

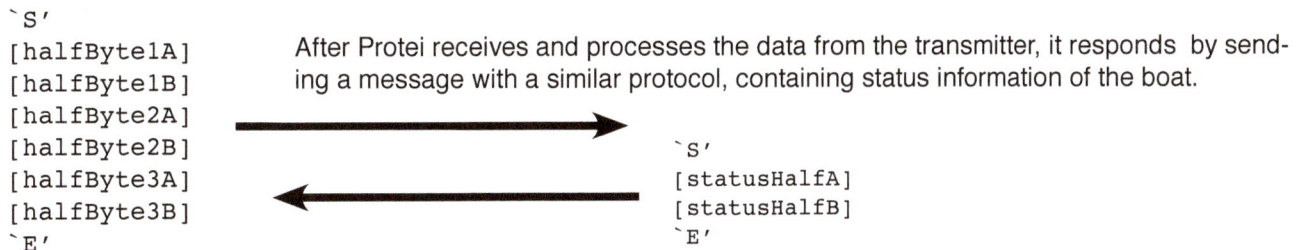

The Xbee Pro 900 (Series 1), used for sending and receiving data.

```
`S'
[halfByte1A]
[halfByte1B]
[halfByte2A]
[halfByte2B]
[halfByte3A]
[halfByte3B]
`E'
```

After Protei receives and processes the data from the transmitter, it responds by sending a message with a similar protocol, containing status information of the boat.

```
`S'
[statusHalfA]
[statusHalfB]
`E'
```

statusHalf, prior to being split into two half bytes and encoded with a Hamming7,4 code, is a bitmap with the following values:

MSB							LSB
N/A	N/A	Stern actuator rotational sensor (HIGH or LOW)	Stern limit switch B	Bow limit switch A	Bow rotation sensor	Bow limit switch B	Stern limit switch A

This information is used by the joystick controller to display six status lights. A seventh status light blinks with each successful radio exchange, and an eighth simply indicates when the controller has power.

The remote controller is a modified hobby RC controller. The electronics have all been replaced with an Arduino and the Xbee described above. Below is an image of the front side of the controller.

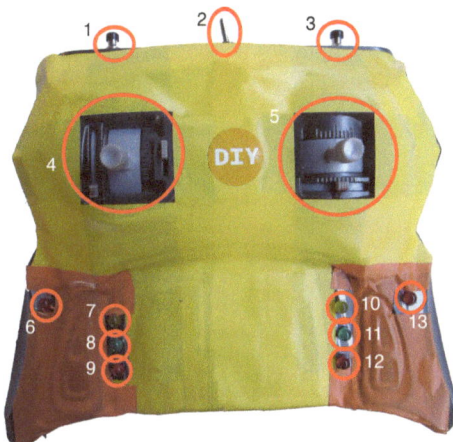

The functionality of the switches and indicators shown on the left are:

1. Sail winch reel in
2. Remote control power
3. Sail winch reel out
4. Bow articulation position
5. Stern articulation position
6. Remote control power light
7. Limit B switch status LED, bow actuator
8. Rotation sensor status LED, bow actuator
9. Limit A switch status LED, bow actuator
10. Limit B switch status LED, stern actuator
11. Rotation sensor status LED, stern actuator
12. Limit A switch status LED, stern actuator
13. Radio link status LED

Microcontroller and connection shield

The primary microcontroller on Protei is the Arduino Mega 2560. We designed and built a custom connection shield for the Arduino. The purpose of this shield is threefold: to contain the logic for controlling the motor driver enable pins, as described in the "Linear Actuators / Sensing" section above; to provide a regulated 3.3V supply with sufficient current for the XBee; and to make easy connections between the motor drivers and sensor inputs, and the pins on the Arduino Mega. The voltage regulator used is an LM1117, with a 22 μF capacitor on the output (required by the LM1117 for stability), as shown on the right:

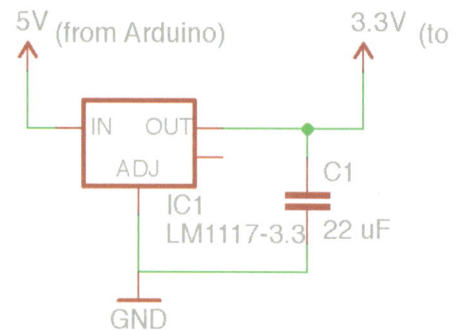

The XBee power supply.

A list of the pin connections are shown below. Pin choices were made by required functionality (PWM, a hardware serial port, etc), then by physical convenience (their location on the Arduino Mega).

Motor Driver 1, Channel A, Enable	42 (Digital input/output)
1A, RPWM	5 (PWM)
1A, LPWM	6 (PWM)
1B, Enable	44 (Digital input/output)
1B, RPWM	4 (PWM)
1B, LPWM	7 (PWM)
2A, Enable	36 (Digital input/output)
2A, RPWM	8 (PWM)
2A, LPWM	9 (PWM)
Bow Actuator, Rotation Sensor	2 (Interrupt 0)
Bow Actuator, Limit A (lower)	38 (Digital input)
Bow Actuator, Limit B (upper)	40 (Digital input)
Stern Actuator, Rotation Sensor	3 (Interrupt 1)
Stern Actuator, Limit A (lower)	46 (Digital input)
Stern Actuator, Limit B (upper)	48 (Digital input)
Winch, Rotation Sensor (currently unused)	21 (Interrupt 2)
Winch, Limit (currently unused)	50 (Digital input)
Xbee, DIN	18 (Serial1)
Xbee, DOUT	19 (Serial1)

A schematic of the layout for this shield is shown below. Red indicates connections directly along the surface of the protoboard, and other colors indicate connections made using wires attached to the protoboard. Note the LM358 in the center left, and the voltage regulator in the upper left.

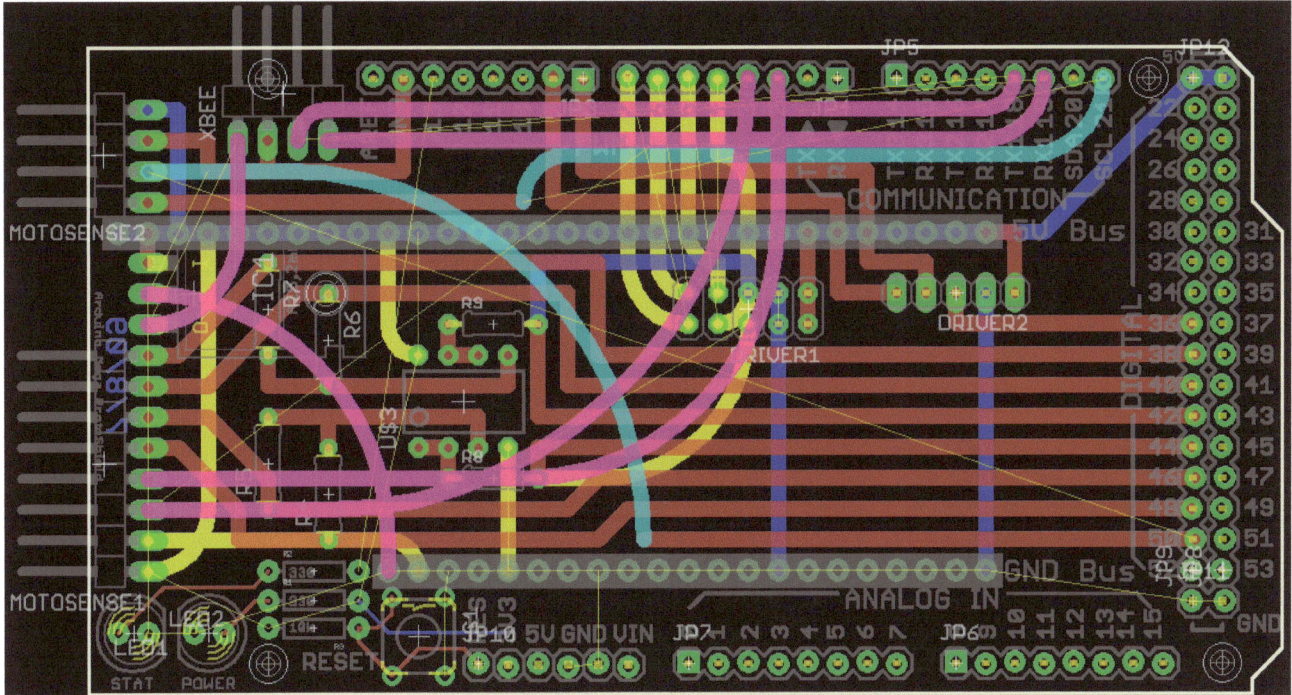

The physical layout of the Arduino Mega shield.

The connector labeled "MOTOSENSE1" has the following connections, from bottom to top: 5V, ground, rotational sensor for actuator 2, rotational sensor for actuator 1, limit switch B for actuator 2, limit A for actuator 2, limit B for actuator 1, limit A for actuator 1. MOTOSENSE2 is not connected in Protei_006, but the connections have been installed for future use. From bottom to top: winch rotational sensor, winch limit switch, ground, 5V. The Xbee connector from left to right is: 3.3V, ground, Xbee DIN, Xbee, DOUT. The motor driver connections are, from left to right, LPWM, RPWM, Enable, ground, 5V.

Two images of each side of the completed Arduino shield are shown below. In the future, it would make sense to have a real printed circuit board professionally fabricated ; however, for the rapid pace of development of Protei_006, it did not.

The assembled Arduino Mega shield.

The Arduino Mega 2560, with its input shield, is mounted with the motor drivers and protection fuses inside a waterproof Pelicase.

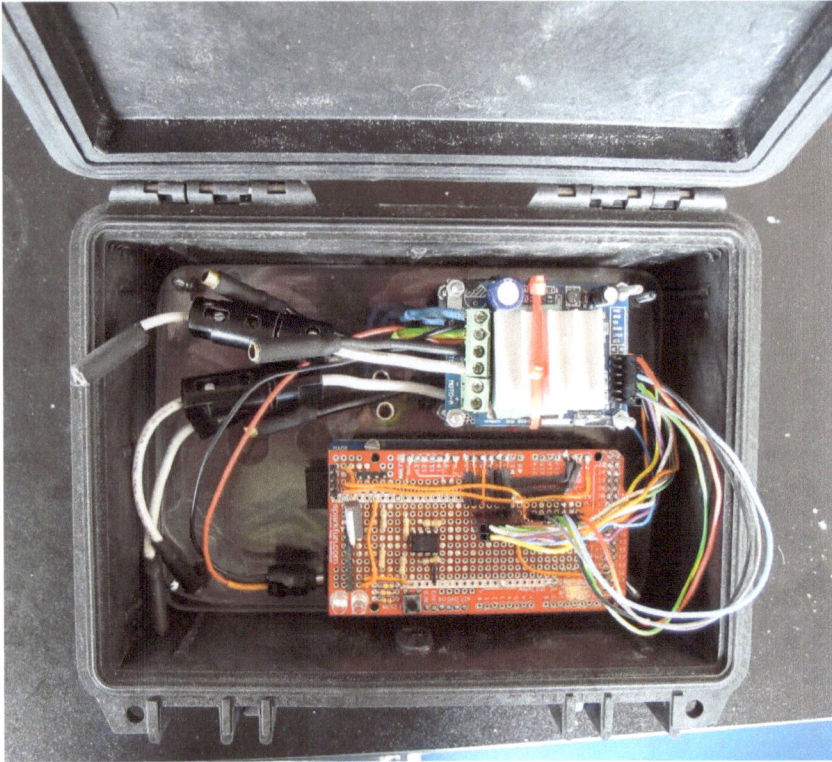

The mounting of the Arduino Mega (with shield), and the two motor drivers (stacked on top of each other) inside the waterproof Pelicase. The colorful cables on the right connect the shield to the motor drivers, and the thick wires on the left provide connections for the motor cables (which enter though a tube on the back of the Pelicase).

Power

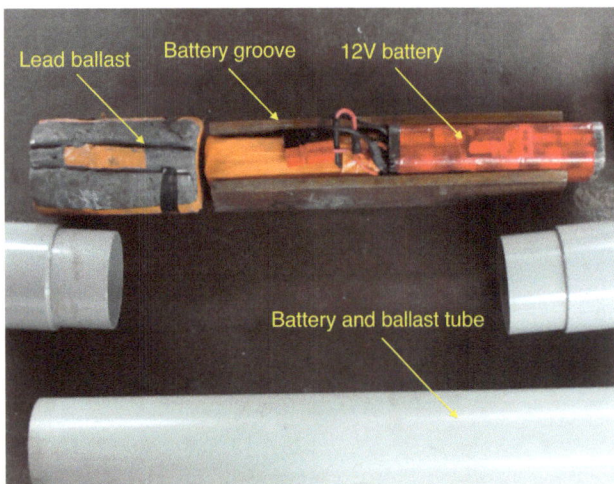

Lead ballast   Battery groove   12V battery

Battery and ballast tube

Power for Protei is supplied by a 14.4V, 9.9Ah nickel metal hydride (NiMH) battery in the keel (see section: Keel Construction, above). Power distribution for this Protei revision is extremely simple — battery power goes through a high current 135A circuit breaker, and from there is distributed to the motor drivers and to the Arduino Mega. Each motor driver output is further protected by a 25A fuse, to protect both the motor driver and the motor in the event of a stall or short circuit. The Arduino provides a 5V power bus, and a 3.3V voltage regulator (described above), provides 3.3V power for the Xbee.

Control firmware

The control process will be summarized below, but for full details, one should examine the source code and its comments directly. The most current firmware version is always available on Protei's git repository.[1]

The control firmware is based around a 20Hz (period of 50ms) control loop. Each loop cycle, the following tasks are performed:

1. Read and process any data the Xbee has received.
2. Send status information back to the transmitter.
3. Each MotorController object runs their main loop:
   a. Calculate the power output necessary for the closed loop actuators.
   b. Send the correct power levels to the two motor driver channels.
4. Send the winch motor power level.
5. Every 4 loops, print any debug information.

There are two important C++ classes used in the control software. The Motor class[2] has references to the 3 pins necessary for motor control, as described above, and to the three necessary input pins (Limit A, Limit B, and the rotation sensor). Through an interrupt (which must, due to the limitations of the Arduino environment, be set up outside of the class), a Motor object can keep track of the number of rotations that it has turned. The move function has some extra logic besides sending the correct PWM values — it also makes sure that the motor comes to a stop before reversing direction (to avoid miscounting rotations), prevents the motor from stopping while the rotation sensor is active (can cause rotation miscounts), prevents very small values from being sent to the motor (to save power), prevents the motor from moving against the limit sensor, and can reassert the Motor driver enable pins after a limit sensor is depressed.

A MotorController class[3] takes a target number of rotations, and a reference to a Motor object, and calculates the power level that should be sent to the motor. If the error (target rotations - current rotations) is less than 2, the motor is told to brake. Otherwise, the motor is told to move at K*error, where K is the gain of the object.

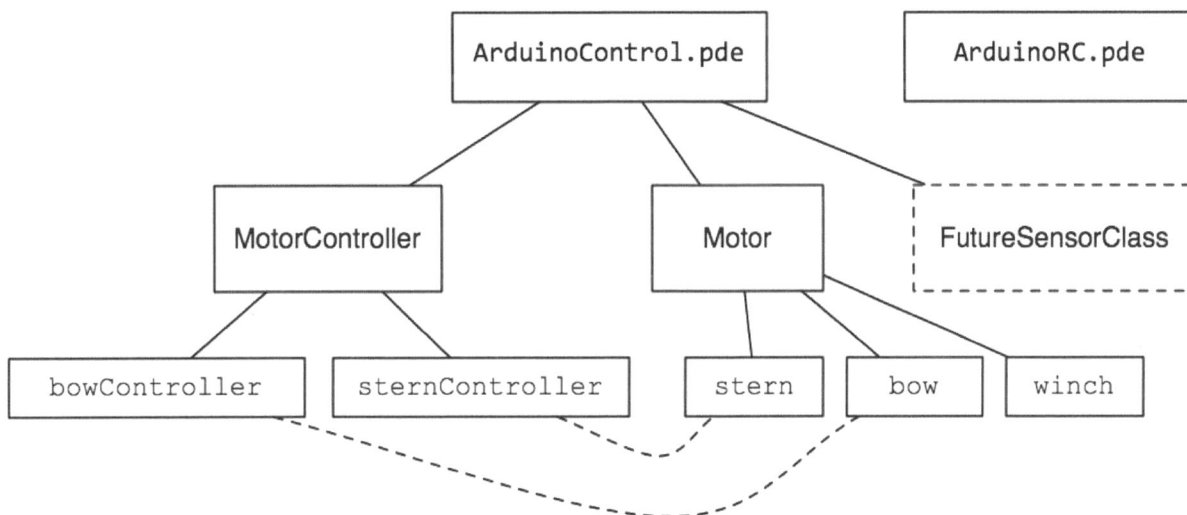

A diagram of the structure of the control firmware and relationships of the classes.

---

1 https://github.com/Protei/Protei-005-6
2 https://github.com/Protei/Protei-005-6/blob/master/ArduinoControl/motor.cpp
3 https://github.com/Protei/Protei-005-6/blob/master/ArduinoControl/motor_controller.cpp

Positional Sensing and GPS

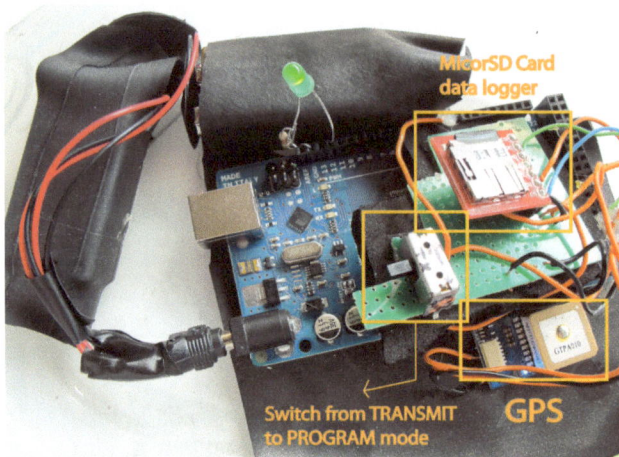

GPS with Arduino and MicroSD card logger

The positional control system is performed using the MediaTek MT3329 GPS. It is an isolated system run off of two 9V batteries, enclosed in a box that fits above the winch box. The GPS transmits incoming NMEA (National Marine Electronics Association) sentences at 10Hz to the Arduino Mega. The NMEA sentences are stored as a text file on a microSD card, using the microSD card shield.

With the GPS data, Protei_006's trip can be visualized on a map simply by uploading the entire text file (or pasting its contents) into GPS Visualizer (http://gpsvisualizer.com).

GPS unit in waterproof box

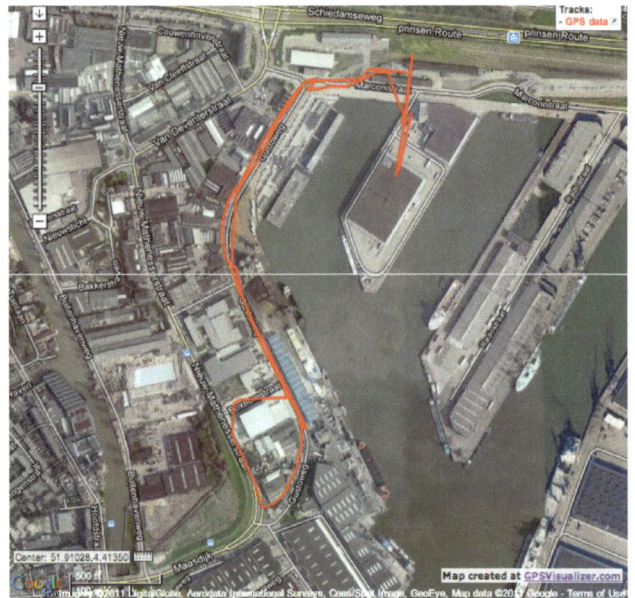

A trip plotted using GPS Visualizer, using the string above

Example of GPS strings:

```
$GPGGA,215341.000,5155.9653,N,00428.0063,E,1,5,4.22,28.1,M,47.,,4
$GPGGA,215342.000,5155.9671,N,00428.0060,E,1,5,4.22,28.0,M,47.1,M,,*6.01
$GPGGA,215343.000,5155.9674,N,00428.0061,E,1,5,4.41,28.2,M,47.1,0E,
$GPGGA,215344.000,5155.9672,N,00428.0059,E,1,5,4.42,29.1,M,47.1,M,,*6A
$GPGGA,215345.000,5155.9683,N,00428.0054,E,1,5,4.22,29.9,M,47.1,M,,*62303R6A
```

NMEA sentences are specified electronically transmitted strings of data, containing global positioning information. For Protei_006, the useful information might include latitude, longitude, course, bearing, speed, time, date, satellite ID's, checksum, and altitude. For more information about NMEA sentences and standards, see http://www.nmea.org/ or http://wiki.openstreetmap.org/wiki/NMEA.

Oil Absorption

Oil characterization

Oil is a complex product due to its wide variety of characteristics. Since Protei application concerns the collection of oil spills at sea, we will focus on crude oils and petroleum products derived from crude oils only.
The main properties of oil are viscosity, density, flash point and API gravity.

• Viscosity is a measure of the resistance of oil to flow, expressed in mPa·s

• Density is the mass of a given volume of oil, expressed in $g·cm^{-3}$. It is the property used by the petroleum industry to define light or heavy crude oils. Density is also important because it indicates whether a particular oil will float or sink in water. As the density of seawater is 1.03 $g·cm^{-3}$, even heavy oils will usually float on it.

• Flash Point is the temperature at which the liquid gives off sufficient vapors to ignite upon exposure to an open flame. A liquid is considered to be flammable if its flash point is less than 60°C.

• API Gravity is based on the density of pure water, which has an arbitrarily assigned API gravity value of 10° (10 degrees). Oils with high densities have low API gravities and vice versa.

Oil type	Viscosity (mPa·s at 15°C)	Density ($g·cm^{-3}$ at 15°C)	Flash Point (°C)	API Gravity (dimensionless)
Gasoline	0.5	0.72	-35	65
Diesel	2	0.84	>62	35
Light Crude	[5—50]	[0.78—0.88]	[-30—30]	[30—50]
Heavy Crude	[50—50,000]	[0.88—1]	[-30—60]	[10—30]
Intermediate Fuel Oil	[1,000—15,000]	[0.94—0.99]	[80—100]	[10—20]
Bunker C Marine fuel Heating fuel	[10,000—15,000]	[0.96—1.04]	>100	[5—15]
Crude Oil Emulsion	[20,000—100,000]	[0.95—1]	>80	[10—15]

Characteristics of different types of oil [Fingas, 2001] and [Wikipedia,Flash Point,2011]

Oil spill characterization

"After an oil spill on water, the oil tends to spread into a slick over the water surface. This is especially true of the lighter products such as gasoline, diesel fuel, and light crude oils, which form very thin slicks. Heavier crudes and Bunker C spread to slicks several millimeters thick.

As a general rule, an oil slick on water spreads relatively quickly immediately after a spill. The outer edges of a typical slick are usually thinner than the inside of the slick at this stage so that the slick may resemble a 'fried egg.' After a day or so of spreading, this effect diminishes.

Winds and currents also spread the oil out and speed up the process. Oil slicks will elongate in the direction of the wind and currents, and as spreading progresses, take on many shapes depending on the driving forces. Oil sheens often precede heavier or thicker oil concentrations. If the winds are high (more than 20 km/h), the sheen may separate from thicker slicks and move downwind." [Fingas, 2001]

The BONN Agreement Oil Appearance Code [BAOAC, 2003] provides a general classification for the quantification of oil on the sea surface, based on its appearance.

Code	Description - Appearance	Layer Thickness Interval (μm)	Liters per km^2
1	Sheen (silvery/grey)	0.04 to 0.30	40 - 300
2	Rainbow	0.30 to 5.0	300 - 5000
3	Metallic	5.0 to 50	5000 - 50,000
4	Discontinuous true oil color	50 to 200	50,000 - 200,000
5	Continuous true oil color	200 to More than 200	200,000 - More than 200,000

BONN Agreement Oil Appearance Code [BAOAC, 2003]

Sorbent boom expected and tested performance

Protei_006 is equipped with standard VanDoClean 6016-B polypropylene booms that can absorb up to 150L of oil according to the manufacturer specifications (see Appendix). They can absorb any type of oil from light diesel to heavy crude, but experience has shown that they are more efficient on light oils, since heavy oils get stuck on the outer surface of the booms [Fingas, 2001]

Reference	Manufacturer	Length (m)	Diameter (m)	Weight (kg)	Absorbency (L)
WB410SN	SpillTech	3.00	0.10	1.60	18.50
6018-A	VandoClean	6.00	0.13	5.00	90.00
WB520SN	SpillTech	6.00	0.13	5.00	56.80
5220	UltraTech	1.80	0.13	2.27	37.90
WB510SN	SpillTechw	3.00	0.13	2.50	28.40
ENV510	SPC - Brady	3.00	0.13	2.20	30.30
6016-B	VandoClean	5.00	0.20	9.00	155.00
SPC816-E	SPC - Brady	5.00	0.20	9.10	114.00
SPC810-E	SPC - Brady	3.00	0.20	5.25	79.00
WB820SN	SpillTech	6.00	0.20	7.25	71.90
WB810SN	SpillTech	3.00	0.20	3.63	36.00
ENV810	SPC - Brady	3.00	0.20	4.30	61.50
BOM510	Dawg Inc.	3.00	0.13	?	30.25
BOM820	Dawg Inc.	3.00	0.20	?	61.50
BOM820G	Dawg Inc.	6.00	0.20	?	123.00
T270	3M	3.00	0.20	5.50	65.00
T270GA	3M	5.00	0.20	8.50	110.00
T280	3M	3.00	0.20	2.75	38.00
T4	3M	1.20	0.08	0.46	3.75
T8	3M	2.40	0.08	0.92	7.50
T12	3M	3.60	0.08	1.38	11.25

Characteristics of different types of sorbent booms (Protei research)

We have also tested a new material from Aeroclay Inc., an American start-up company dedicated to commercializing an advanced material aerogel technology known as AeroClay®.

The following results have been obtained by iterating a saturating-and-squeezing process on two samples : one of Aeroclay and one of standard polypropylene oil absorbent (VanDoClean 6016-B)

The dry weight of the sample is referred as S_0:

S_0 = 18g for the AeroClay® sample (16g of material + 2g for the surrounding pillow)

S_0 = 14g for the polypropylene sample

Cycle #	Total oil absorbed (O_S)	Net oil remaining (O_N)	Ratio O_S/S_0
1	168 g	82 g	9.33
2	164 g	86 g	9.11
3	162 g	90 g	9
4	142 g	80 g	7.89
5	148 g	80 g	8.22
6	148 g	82 g	8.22

AeroClay® saturation results

Cycle #	Total oil absorbed (O_S)	Net oil remaining (O_N)	Ratio O_S/S_0
1	102 g	62 g	7.29
2	94 g	58 g	6.71
3	88 g	52 g	6.29

Polypropylene saturation results

After 3 cycles, the polypropylene sample was so damaged by the squeezing process that it was considered to be unusable.

On the other hand, the AeroClay® sample remained functional even after the sixth iteration.

Our conclusion is that AeroClay® material seems very promising compared to existing industrial sorbent technology because :

1. The global absorbency efficiency of AeroClay® (measured by the O_S/S_0 ratio) is superior to that of polypropylene.
2. It has been shown to be reusable even after a multiple squeezing cycles, unlike polypropylene.
3. From the preliminary results, it seems that the squeezed oil can be reused, whereas the current sorbent technologies usually end up burned or buried in landfills. [Fingas, 2001]

However, no large-scale test results are available yet, so we cannot guarantee the performance of AeroClay® over a full-scale recovery operation. Nevertheless, tests indicate that AeroClay® might represent a good solution for Protei's oil absorbing system; an AeroClay® boom could last much longer than a traditional one.

Deployment of Protei for oil spill collection

Protei_006 is designed to tow a sorbent boom that can collect light oils with the following typical characteristics :
*   Density between 0.7 and 0.9 g•cm^{-3} (API between 25 and 70)
*   Flash Point above 60°C (Protei cannot handle flammable products)
*   Viscosity under 50 mPa•s

Protei_006 is completely remote-controlled and cannot stay at sea for long, due to its battery limitations. Future versions of Protei should be able to sail autonomously for long periods of time — maybe weeks — and will therefore be able to collect oil on large areas.

Furthermore, existing technologies such as skimmers are already very efficient on thick slicks (codes 4 and 5, see BAOAC table), and Protei is not intended to replace them. Instead, Protei is aimed at thin slicks (codes 1 to 3), on which existing technologies are not efficient currently.

Possible scenario : discharges from offshore oil installations

Motivations

One of the possible scenarios is the cleaning of water discharges released by offshore platforms. These still contain a fraction of oil and can cover very large surfaces areas in remote places.

Although they meet the maximum oil content in water recommended by the Oslo-Paris convention (30g/mL) [OSPAR, 2001], these leaks still represent a significant amount of uncontrolled oil poured into the oceans. Protei could provide a cost-effective alternative to existing technologies on this particular scenario.

Static recovery performance calculation

Oil films from water produced by the offshore platforms were observed by SINTEF [SINTEF, 2011] :

- Up to 40 km from the platform
- Sheen / rainbow / metallic colors
- Within the recommended limits regarding oil content (< 30 ppm) [OSPAR, 2001]

Aerial Estimates:
- Length : 1000m
- Width : 30 m
- Estimated area : 0.04 km²
- 50 % Sheen
- 50 % Rainbow

Total estimated volume:
- BAOAC : 7 L (low) - 106 L (high)

Releases:
- 22.000 m³ / day
- 25 ppm oil
- 600L oil / day
- 30L oil / hour

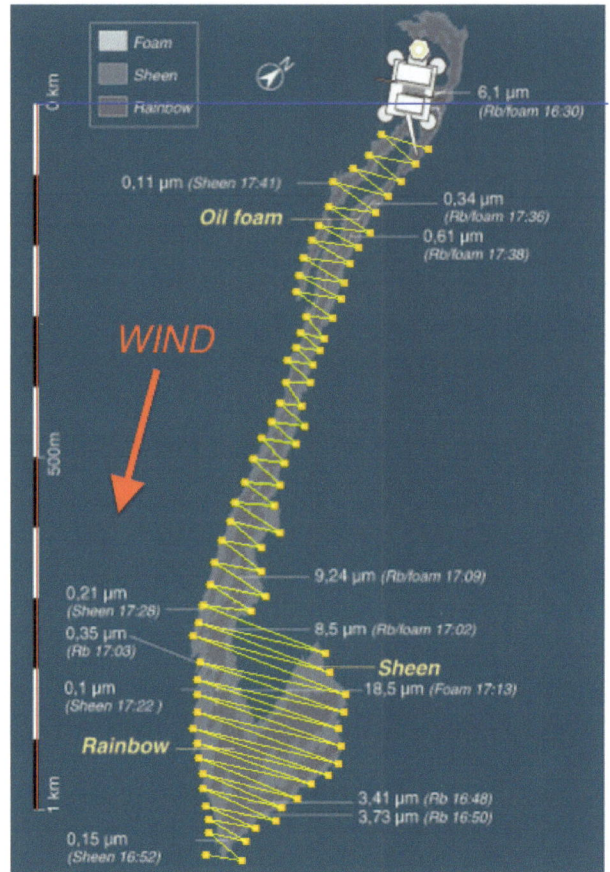

Upwind sailing path to comb the oil slick

In a static calculation, the discharge is seen at a fixed instant: the oil slick has a constant volume. The objective is to clean this static oil slick in one go, with one boom. Is that possible?

Assumptions :
1. Discharge rate : 0L oil / hour (fixed instant)
2. Initial volume : 106L (upper bound of the BAOAC estimated volume)
3. Wind coming from the North at a constant speed
4. Protei sailing at 1.5 knots upwind, constantly (even when tacking), for maximum sorbent efficiency
5. Sorbent boom : VanDoClean 6016-B with a 150L capacity

Calculation :
1. An upwind sailing path combing the whole static oil slick is designed (see the yellow line on above picture)
- 82 points = 82 tacks
- 7920 m long path
- Trip total duration at 1.5 knots : 2hrs 50min

2. This means that Protei needs to collect 106L of oil in 2hrs 50min. Since the absorption capacity of the boom is 150L, it is sufficient. The needed absorption rate is then 106L in 2hrs 50 min = 37.2L of oil/hour to clean the whole oil slick in one go.

This performance is considered optimal because of the static discharge assumption and the constant sailing speed assumption. When tacking, the average sail boat slows down. With its articulated hull, Protei is meant to be able to tack without losing speed.

The absorption rate of the sorbent boom needs to be measured to see how it compares to this optimal performance. If it compares favorably (higher than 37.2 L/hour), then it is theoretically possible to clean the whole oil slick in one go with one boom.

However, the spill we are looking at is not a static one. It is a continuous leak from the platform discharge point, at a constant flow rate of 30L/hour, which represents 85L in 2hrs 50min. A dynamic calculation would be necessary. Here again, the absorption rate of the sorbent boom needs to be measured and compared to the figure of 30L of oil/hour. If it is more than 30L oil/hour, then it is possible.

The challenge with measuring the absorption rate is that the value mostly depends on parameters such as the oil viscosity (it is harder to absorb a viscous product) and the thickness of the oil slick (if the slick is too thin, there is nothing to absorb). Therefore the measurement is very difficult to replicate and can only be obtained with a fixed set of parameters. In the real world the thickness of the oil slick changes from the border of the slick to its center, being usually thinner at the border [Fingas,2001].

Compatibility with existing oil spill contingency plans

Existing response plans for oil spills have been thoroughly tested and improved through the past decades in countries where major oil spills happened in the past, with specific procedures, proper equipment and trained personnel. For an example, see the American National Oil and Hazardous Substances Pollution Contingency Plan [US EPA 2011].

Protei is not intended to replace any of the technologies currently in use for such response plans. However, Protei might provide a good alternative if the conditions make it is too hazardous for human exposure, on which the existing contingency plans rely. These conditions might include extreme weather, hazardous chemicals, local terrorism or unstable political situation.

Protei_007
Future developments

Known issues and proposed improvements

The current model of Protei is somewhat balanced, maneuverable (by adjusting the trim of the sail and the curvature of the hull), responds to RC control, and heels in the wind while staying upright. Because of its round, serpent-shaped hull, its heavy ballast, and its low center of gravity, in high winds and unstable conditions, Protei is intended to lay flat on the water until conditions improve and it can right itself.

However, there are improvements for the build of Protei_007, to optimize its behavior, robustness, and maneuverability. Some improvements include the following:

Protei must be able to successfully tack back and forth across the wind - this will be achieved by better control over the hull articulation (the current motors have been replaced with more powerful ones). More tests must be performed towing the oil boom as it gets saturated with oil, as well as in rough oceanic and meteorological conditions. Eventually, Protei will need to be able to obey RC commands from longer distances than the XBees can provide.

The biggest improvement for Protei_007 will be the implementation of a water-proofed unit in which all the electromechanical equipment lives (rather than separate compartment for each component). With one centralized space, it will be easier to find the source of a leak, as well as to develop an automated system to get rid of the water, with a water sensor and a bilge pump.

With regards to the skin, we did not research the chemical aspects of the EPDM-Foam yet, including how it will be affected when in contact with crude oil. Although this material provides flexibility and buoyancy, it rips or tears easily when in contact with a sharp object. Additionally, the skeleton of the hull is composed of PVC, and due to its toxic components, a better material should be used.

A better system for launching and retrieving the vessel should also be implemented, for example with the use of a small crane or pulley mechanism. For the first test of Protei_006, the boat was lowered vertically 2m into the water. When retrieving it, the EPDM-Foam skin tore (about 2cm) as it rubbed along the wall.

The use of mechanical limit switches on the linear actuators can be improved upon, because mechanical limit switches have inherent reliability issues. Switching to an optical switch or sensor might be more reliable.

Further testing needs to be performed to determine if Protei_006 is truly self-righting, how robust it is in extreme weather conditions, and how it behaves when towing a saturated oil boom.

Evolution

As the design of the individual Protei vessel improves, further versions will develop towards the behavior of multiple vessels, making up a swarm of sailing drones. As we progress, the vessel will move towards energetic autonomy, possessing sensing and decision-making skills, and eventually will evolve from a centralised swarm control to a decentralised peer-to-peer autonomy.

The evolution of Protei is divided into six levels of networked systems, each building towards a more autonomous agent (the individual boat) with the capability of complex interactions (amongst the individual agents, and between multiple networks of agents). Eventually, Protei will be a self-organizing, multi-platformed network, with web-interface capabilities, to enable over-riding a unit (or units) of the swarm, whilst not disturbing the network as a whole. This current Protei model is between level 1 and 2: It is a radio controlled vessel, needing a human controller nearby (within RF range) to steer and control it. It is beginning to log environmental and positional data about itself and its journey (by storing GPS information).

The multiple Platforms of Protei
Individual Vessels (Platforms 1, 2, and 3):

1. Radio controlled with one user

The first platform for Protei is one single remotely controlled boat, requiring 1 user nearby. Control includes steering by directing the articulation of the hull, as well as determining the trim of the sail.

Protei_001 through Protei_006 are of this platform.

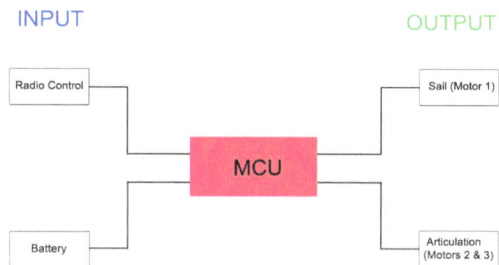

2. Radio controlled + sensing (semi-autonomous)

The second Protei platform is a radio-controlled vessel, capable of sensing and storing data about its surroundings. It requires 1 user to control it remotely, but it has built in positional and environmental sensors (including a GPS, compass, gyroscope, accelerometer and anemometer).

Protei_006 is between this platform and the first one, in that Protei_006 logs GPS data.

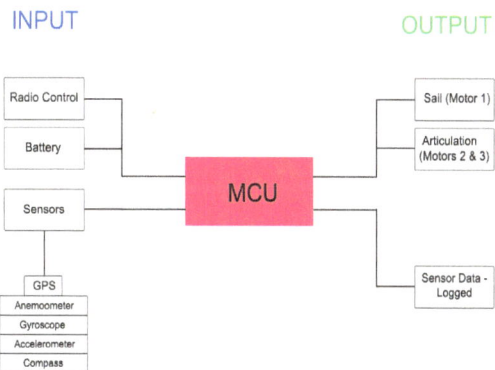

3. Autonomous

The third platform of Protei is a fully autonomous unit, that takes in data about itself and its surroundings, and makes decisions about appropriate behavior and actions to take. This platform no longer needs a human operator. The vessel sails independently, follows way points, rights itself, tacks across the wind, and adjusts its behavior due to changing environmental stimuli.

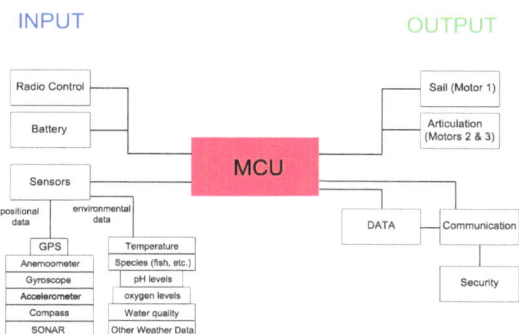

Multiple Units making up a complex system of vessels (4, 5, and 6):

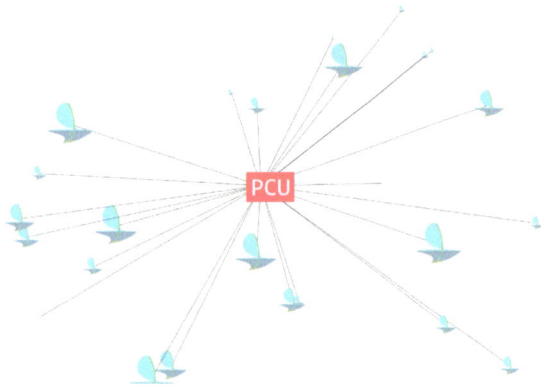

Master-slave model: a PCU governs slave behavior, and the slaves communicate back to the master

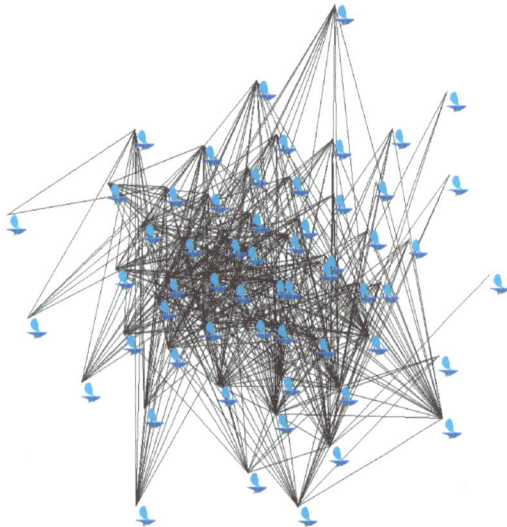

Complex system behavior; Peer-to-peer communication; Input and output information sent through individuals, the network, and sub-networks)

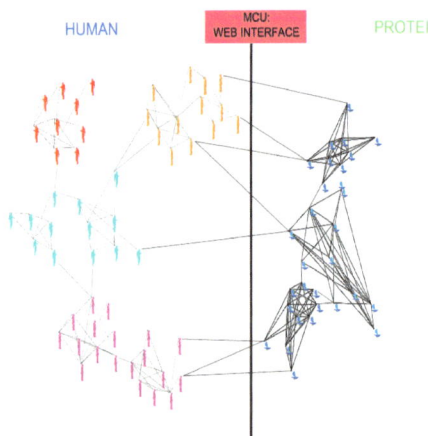

A multi-level platform of networks and subnetworks of vessels; control is possible via a web interface

4. Swarm: Centralized

The fourth evolution of Protei is a network of multiple individual vessels that obey commands from one centralized PCU master which controls group behavior. There is no communication between vessels. Each individual vessel does not need computational capacity for steering because the PCU sends instructions: This requires minimal network optimization, and the group moves as one unit. If one unit gets too far from the master, it will get lost.

5. Swarm: Decentralized

The fifth evolution of Protei is a non-linear, complex, self organizing system, consisting of subsystems and individuals. Each individual possesses intelligence to compute its position in the environment, relative to nearby individuals, and to the entire group. A suite of behavioral characteristics drive each organism's actions and interactions. This platform requires that each individual makes predictions about environmental characteristics, such as wind and current, through receipt and transmission of peer-peer information. At this level, the control system of Protei moves beyond Arduino, to a more complex, multi-level operating system, such as Android.

6. Multi-Platform

The most complex level of the Protei Platform brings back the human interface. This level includes multiple independent networks of vessels, with an MCU that is an interface between humans and the networks of vessels. This allows for a web platform for individuals (humans) to take control of single or multiple vessels to override its control mechanisms, without disrupting the network as a whole. This allows for a robust, adaptable system, with many behavioral levels, including groups, sub-groups, and individuals. This adaptive, emergent system is dependent on complex algorithms, machine learning and network optimization protocols.

Applications

Protei can eventually be appropriated for other purposes, such as cleaning other chemical pollutants and material waste from the water, as well as collecting samples for ocean research. As we move towards Protei_007 and beyond, we are going towards a more modular design, made from recycled materials, so that the vessel is an adaptable platform for multiple purposes. Some other applications for Protei may include:

1. Plastic collection

In the North Pacific Gyre, Protei could be equipped with a fine mesh net and used to collect plastic waste. It has been estimated that 15,000,000 square kilometres contain unusual accumulation of small suspended plastic particles (Marks, Kathy 2008). As plastic waste accumulates in the gyres, the combined effect of UV light, the mechanical action of waves, the saltiness, and the acidity breaks it down. At such a small size, animals mistake the plastic for plankton and ingest it in quantities. Some animals die of accumulation and obstruction of their digestive system; Some go up the food chain and eventually end up in our plates where they return us the toxic we produced. Even if we stopped producing plastic now, it will take decades for the plastic to break down into smaller bits and accumulate in the gyre. There is no point in developping a very fast plastic recovery system, but rather a system that captures the plastic at the same rate as the one at which the current circulates, hence working with the surface currents and dominant winds. Very few technologies have been developed to solve this issue and Protei may be able to contribute since it may be capable of performing the very repetitive task of plastic collection over immense areas. One other key to the issue may be to re-qualify the plastic at sea not as a form of pollution but as a new resource, a source of energy, construction material, and therefore, profit.

2. Radioactivity monitoring

Following the Fukushima nuclear accident, the International Atomic Energy Agency (IAEA) has been compiling and analyzing sea-water sample data collected by TEPCO (the operator of the damaged plant) and MEXT, two organizations with offshore stations near the site of the nuclear power station (IAEA slides, 2011). The dominant winds in the area are pushing the radioactive particles in the air towards the pacific. At the worst of the nuclear crisis, seawater was used to cool down the reactor, which was placed directly back in the ocean. The information collected at an offshore fixed site does not provide the crucial high levels of radiation close to the origin of the pollution, and many points fur-

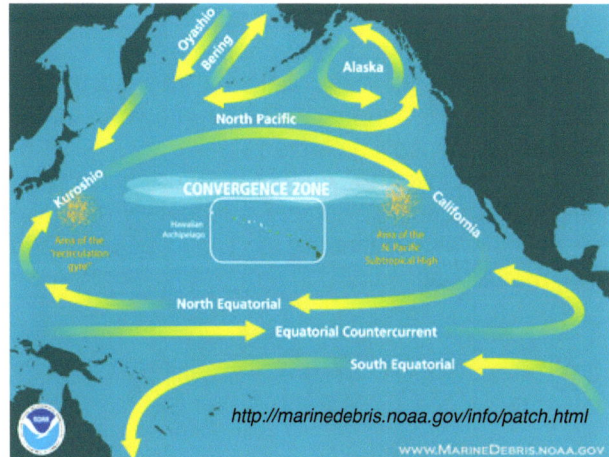

Ocean currents in the Pacific Ocean

Sample of plastic garbage
Algalita Foundation.

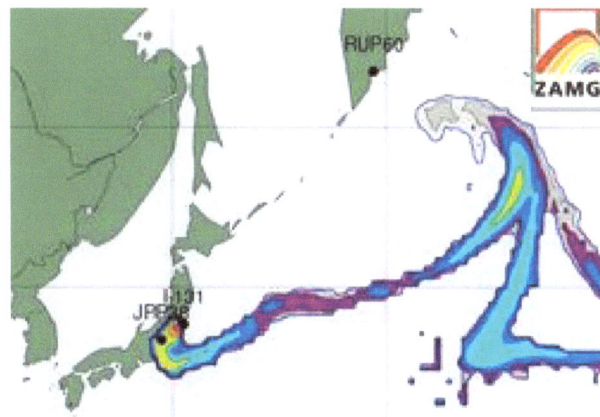

Radioactive plume over Japan

ther out at sea have no corresponding data. We cannot make manned measurements of high levels of radiation - Drones are needed for that. Protei technology can be implemented for this type of marine environmental monitoring and mapping, as an alternative to sending humans to such potentially hazardous areas.

3. Climate and oceanography research

For example, in pirate-infested waters of the Indian Ocean, climate scientists aim to collect data about salinity, water temperatures, and weather systems in order to map global networks of deep oceans. Because of piracy threats, they are restricted from accessing about a quarter of the Indian Ocean (Fogarty, 2011). Alternatively, robotic, unmanned vessels such as Protei, can be deployed to conduct measurements from such politically unstable areas. Additionally, Protei, like other robotic vessels that are starting to be launched in the India Ocean, can aid in anti-piracy efforts.

4. Water sampling and quality monitoring in coastal areas

Unmanned vessels, such as Protei, could aid in measuring concentrations of microscopic toxins such as PCBs. Specifically, the European Commission is rolling out an extensive water quality assessment program (Water Directive 2020), for which they need extensive samples. Protei can provide a cheap and reliable means to collect biological samples and to measure other things such as water quality.

4. Monitoring marine protected areas

Over the last generation, advancements in digital technology and naval engineering have allowed fishing fleets to scale to enormous size and extract catch with great effectiveness. As a result, our precious oceans currently face an unprecedented threat of overfishing and ecosystem destruction. A prominent assessment was taken in 2009 and showed that, as a result of overfishing, 63% of assessed fish stocks worldwide still require rebuilding [Worm & Hilborn, 2009]. Without effective monitoring and control, the future of our delicate fish stocks is in danger. Both overfishing and Illegal, Unregulated and Unreported (IUU) fishing help commercial fishing to be the single greatest pressure on our remote marine ecosystems [SERMA]. For all but the wealthiest countries, the many marine areas are too remote or too difficult to protect using current monitoring approaches.

Current worldwide economic losses as a result of IUU

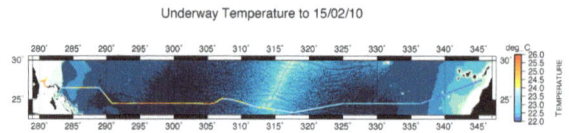

Underway Temperature to 15/02/10

P. Keen

Deployment of oceanographic instrumentation

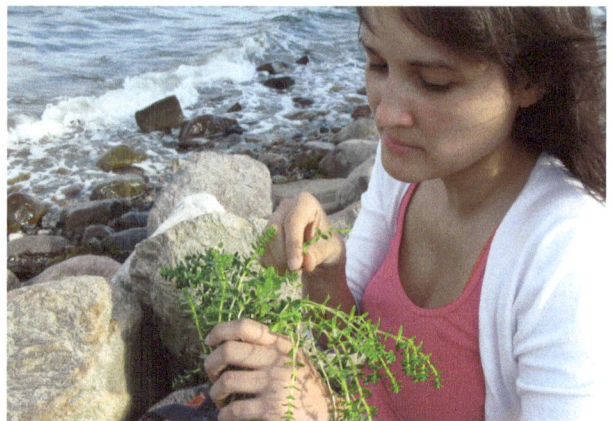

Algae sampling

PW Keen Marine Consultancy

fishing have been estimated to be as much as US$23.5 billion annually, which is on the order of 22% of total fisheries production [Agnew, 2009]. The majority of IUU fishing typically targets developing nations, exploiting the lack of protection resources to rob the poorest people on the planet. An estimated one billion people [FAO, 2000] rely on the oceans as a primary source of dietary protein, which makes the issue not only an economic and environmental concern, but also a human rights and food security one. What is needed is a better way to watch over our oceans.

Protei, under remote operation from a ground site or autonomous pre-programmed path, can help by monitoring these remote marine areas or places that have been designated

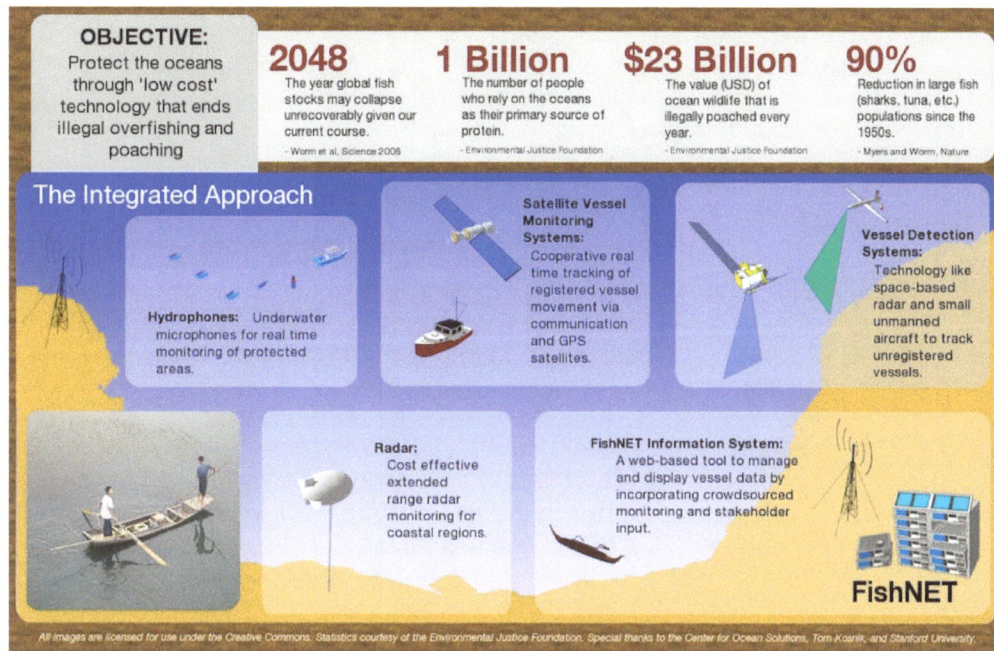

OBJECTIVE:
Protect the oceans through 'low cost' technology that ends illegal overfishing and poaching

2048
The year global fish stocks may collapse unrecoverably given our current course.
- Worm et al, Science 2006

1 Billion
The number of people who rely on the oceans as their primary source of protein.
- Environmental Justice Foundation

$23 Billion
The value (USD) of ocean wildlife that is illegally poached every year.
- Environmental Justice Foundation

90%
Reduction in large fish (sharks, tuna, etc.) populations since the 1950s.
- Myers and Worm, Nature

The Integrated Approach

Hydrophones: Underwater microphones for real time monitoring of protected areas.

Satellite Vessel Monitoring Systems: Cooperative real time tracking of registered vessel movement via communication and GPS satellites.

Vessel Detection Systems: Technology like space-based radar and small unmanned aircraft to track unregistered vessels.

Radar: Cost effective extended range radar monitoring for coastal regions.

FishNET Information System: A web-based tool to manage and display vessel data by incorporating crowdsourced monitoring and stakeholder input.

FishNET

All images are licensed for use under the Creative Commons. Statistics courtesy of the Environmental Justice Foundation. Special thanks to the Center for Ocean Solutions, Tom Kosnik, and Stanford University.

as protected areas. Many of the patrol vessels for developing nations remain at port as a result of the lack of resources to fuel them. Enforcement of the marine protected areas in many parts of the world is far too expensive for all but the wealthiest nations. The South Pacific is a perfect example of a region where robotic vessels can fill the gap. Protei can patrol the oceans; provide real-time feedback and marine situational awareness to mitigate IUU fishing.

These vessels can be incorporated into an integrated protection network to expand the capacity of governments to monitor and control their EEZs. FishNET, a network of connected technologies that aims to stop illegal fishing, plans to incorporate Protei into its suite of available technologies. With this input from these vessels, the integrated information database would be much stronger and capable as Protei is the perfect tool to extend the range of protection. Accompanying unmanned aircraft, underwater acoustic hydrophones, blimp/air-based relays, and satellite technologies (see image), Protei and FishNET can work seamlessly for effective fisheries and ecosystem management.

Frequently asked questions

How can I find out information about Protei that is not in this handbook, and stay updated about the progress?
- Our website, https://sites.google.com/a/opensailing.net/protei, is a great place to start.
- E-mail us! (contact@protei.org). Request to be added to the "insider" mailing list.
- Join Open_Sailing group on Facebook: http://tinyurl.com/open-sailing-facebook
- Check out our photos at http://www.flickr.com/groups/protei/

Who controls Protei?
Currently, the Protei team has "driven" Protei_006 on the water. The idea is that it can be remotely controlled from the shore or a nearby vessel, although hopefully control can be long-range. Eventually, we hope to build a Protei that has sensors in order to navigate semi-autonomously, and then a fleet of Proteis that would communicate and operate as a network. However, if Protei_006 was developed to be more robust, it can be controlled by fishermen, first responders, gamers, sailing enthusiasts, scientists and hobbyists alike.

Who will build a fleet of many Protei vessels?
Because Protei is an open source project, its success as a fleet depends on ongoing contributions to its design and functionality, and its construction and implementation on a large scale. It must be a widespread effort for it to be a network of vessels, so our hope is that many people will build and use Protei.

What is Open_Sailing and how is it related to Protei?
Open_Sailing is a growing international community with the goal of developing open-source technologies to explore, study, and preserve oceans. Protei is one of many Open_Sailing projects.

How did Protei start?
Cesar Harada[1] conceived of Protei one night in his kitchen while he was working at MIT with Sey Min[2]. They came up together with the name. From the start, it was an open-source project. In its early stages Protei was facilitated by LA Bucket Brigade[3], Earl Scioneaux III, Jessica Rohloff and Suzette Toledano[4] in New Orleans. The TED network[5], including Nate Mook[6] and David Troy[7] immensely helped Protei take over. Quickly the project became a collaboration between Open_Sailing[8] and randomwalks (Protei_005), but soon V2_ Institute for the Unstable Media[9] in Rotterdam decided to support the development of Protei. Protei is now a team effort, joined by the design collective Amorphica[10] and many independent architects, designers, engineers, inventors, researchers, students, retired geniuses, makers, DIYers and Open Hardware advocates. Protei has since grown into international collaboration.

What do you do with the oil?
Discarding the oil is not in the scope of Protei functionality. However, the ideal goal is that there will be a recycling plant on land for the oil.

---

1   http://www.cesarharada.com
2   http://randomwalks.org/
3   http://www.labucketbrigade.org/
4   http://www.suzettebecker.com/
5   http://www.ted.com/
6   http://betanews.com/author/nate
7   http://davetroy.com/
8   http://opensailing.net/
9   http://www.v2.nl/
10   http://www.amorphica.com/

How much does it cost to make one Protei? Can I buy Protei?

Every prototype has its own cost, depending on the materials required and its capabilities. Including tools and other necessary items for construction, Protei_006 was allotted $10,000. Protei, at the moment, is not a stable release, but could be developed into one.

Does Protei_006 fit in a van ? Yes it does.

Appendices

Parts list

Subsystem	Components	Manufacturer	Price	Reference
Hull	PVC Tubing 24, 26, 22, 16, 19, 12mm Ø			
	3.5 mm thick Plywood			
	Polystyrene Foam			
	Bowden Cables, Connectors and Housing			
	Springs			
Sail and Mast	Nylon (Sail Material)			
	Windsurfing Carbon Mast	Prolimit STX C60 RDM	$242	http://www.surfladle.co.uk/products/product-detail.php?PID=2354
	Windsurfing Aluminum Boom	Neilpryde 230-275 X5	$143	http://www.surfspullen.nl/windsurf/gieken/neil-pryde-x5-230-275-alu-2006/3100/2776
	Aluminum Stock (for couplings)			
Keel	Wooden Plates			
	Garden hose (for cable routing)			
	EPDM-Foam	Panacell		http://www.panacell.de/cms/material/zellkautschuk.html
	PVC Tubing 22mm Ø			
	Rubber Inner Tubes			
Skin	EPDM-Foam			
	Rubber Inner tubes			
	Nylon Ropes			
	Rubber Glue			
Linear Actuator	PVC Tubing 63mm Ø and Connectors			
	Universal Joints			
	Hose clamps			
	Wood			
	Copper Tubing			
	Aluminum Stock			
	Bowden Cables, Connectors and Housing			
	Bosch DC Motor and Gearbox extracted from hand drill	Bosch	$349.96	http://www.amazon.com/Bosch-34614-14-4-Volt-2-Inch-Compact/dp/B000VZP5W2

Ballast	Lead			
	PVC Tubing 95mm Ø			
	Steel Plates			
Winch	Plexiglass			
	Wood			
	Aluminum Tubing			
	PVC Tubing			
	Windshield Wiper Motor	Varies	Varies	
Electronics	Pelican Case			
	Arduino Mega 2560	Arduino	$64.95	http://www.sparkfun.com/products/9949
	Arduino Mega Shield	Sparkfun	$17.95	http://www.sparkfun.com/products/9346
	50A Freescale H-Bridge Motor Driver	Freescale	$107.80	http://www.elechouse.com/elechouse/index.php?main_page=product_info&products_id=667
	Misc. Electrical Components (wire, resistors etc)			See electrical schematics for more info
	14.4V 10Ah Ni MH Battery	BatterySpace	$145.50	http://www.batteryspace.com/mhbattery144v10ah144wh forportabledevices.aspx
	Xbee Pro 900 Series 1	Digi	$85.90	http://www.sparkfun.com/products/9097
	Remote Controller			
Miscellaneous	Hardware (Screws, Washers, Rivets , nuts etc)			

Nautical nomenclature

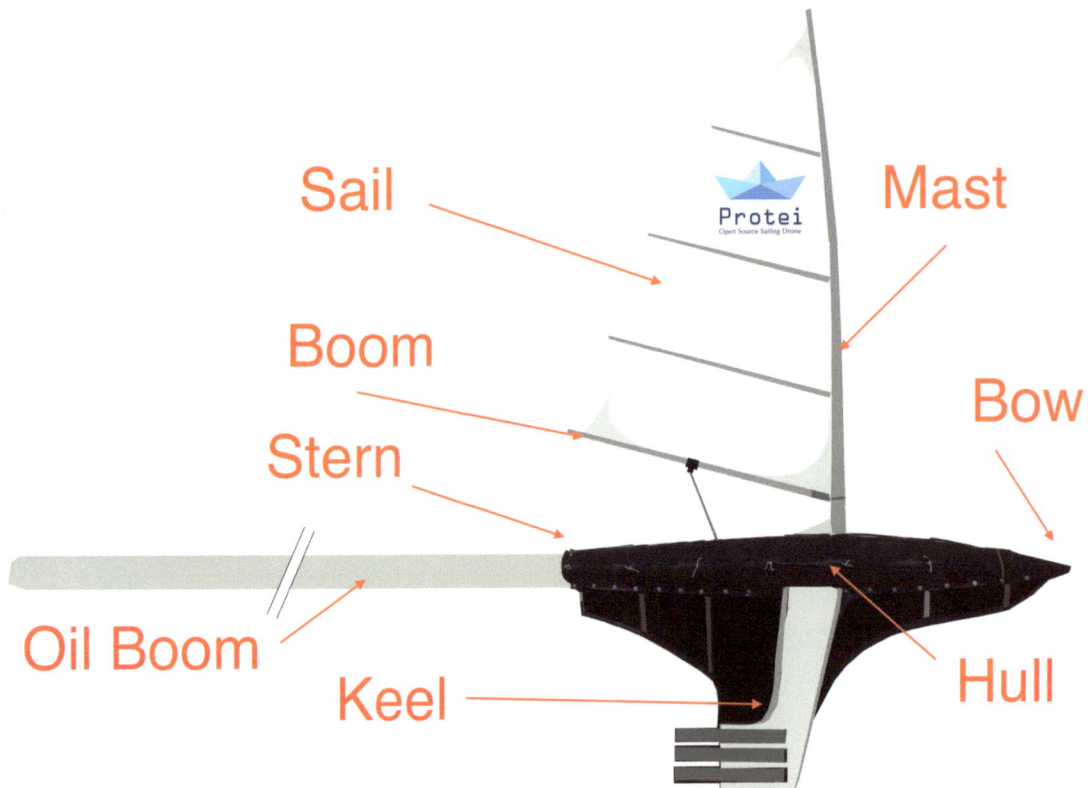

Sail

Mast

Boom

Bow

Stern

Oil Boom

Keel

Hull

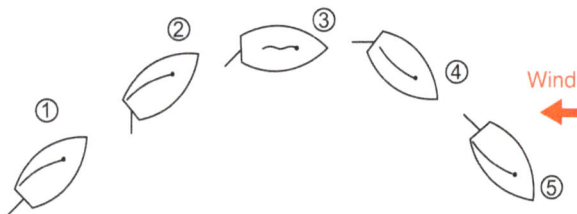

Schematic of Tacking (Kangel, 2011)

Wind

Bow: The forward-most part of the hull.

Stern: The back part of the hull.

Starboard: The right side of the boat when facing forward.

Port: The left side of the boat when facing forward.

Tack: Sailing maneuver that involves turning the boat so that its bow crosses the wind.

Bulkhead: A vertical wall in a ship.

Amidships (or midships): Positioned at the middle of the boat (could refer to both middle in the bow-stern line or starboard-port line).

Protei community

From left to right : Sebastian Neitsch, Qiuyang Zhou, Gabriella Levine, Logan Williams, Sebastian Müllauer, Roberto J. Meléndez, Toni Nottebohm, Piem Wirtz, Cesar Harada, Photo by Willem Plet.

Protei_team

Cesar Minoru Harada
Qiuyang Zhou
Sebastian Müllauer
François de la Taste

Logan Williams
Gabriella Levine
Piem Wirtz
Roberto J. Meléndez

Peter Keen
Etienne Gernez
Toni Nottebohm

Collaborators

Jiskar Schmitz
Henrik Rudstrøm
Sebastian Neitsch
Fiona Crabbie
Boris Debackere
Kasia Molga
Pinar Temiz
Julia Cerrud
Giulia Garbin
Alexia Boiteau
Isidora Markou
Hunter Daniel

Roman Yablonski
Aaron Gutierrez
Dr. Ing. Gonzalo Tampier
Javier Henríquez Quezada
Pr. Zenon Chaczko
Philippe Noury
Joshua Updyke
Simon de Bakker
Pr. Jennifer Gabrys
Dr. Dominic Muren
Dr. Sarah Jane Pell
Narito Harada

Matthew Lippincott
Molly Danielson
Dillin Harr
Mario Saenz
Kisoon Olm
Sunghun
Jieun Yoo
Tyler Survant
Daniel Henneberger
Maia Anthea Marinelli
Daniel Henneberger
Shah Selbe

Partners

Open_Sailing (*www.opensailing.net*)
V2_ Institute for the Unstable Media (*www.v2.nl*)
DNV (*www.dnv.com*)
Amorphica (*www.amorphica.com*)
RandomWalks (*www.randomwalks.org*)
Hofman and Zonen (*www.florentijnhofman.nl/dev/*)
TED (*www.ted.com*)
TEDxMidAtlantic (*www.tedxmidatlantic.com*)
Kaag Watersport Academy (*www.dekaag.nl*)
Creole.com (*www.creole.com*)
Goldsmiths University (*www.gold.ac.uk*)
MIT Public Service Center (*web.mit.edu/mitpsc/*)
HTS Consulting (*www.htsconsulting.com*)
LA Bucket Team Brigade (*www.labucketbrigade.org*)
Suzette Toledano Becker (**www.suzettebecker.com**)

Supporters

All our 300+ Kickstarter backers !
Florentijn Hofman and Kim Engbers
Shannon Dosemagen
Marnix de Nijs
Diana Wieser
Dana Braff
Jasper van Maede
Jean-Grégoire Kherian
Earl Scionneaux III
Jessica Rohloff
Nae Morita
Nick Kaufmann
Michel Van Dartel

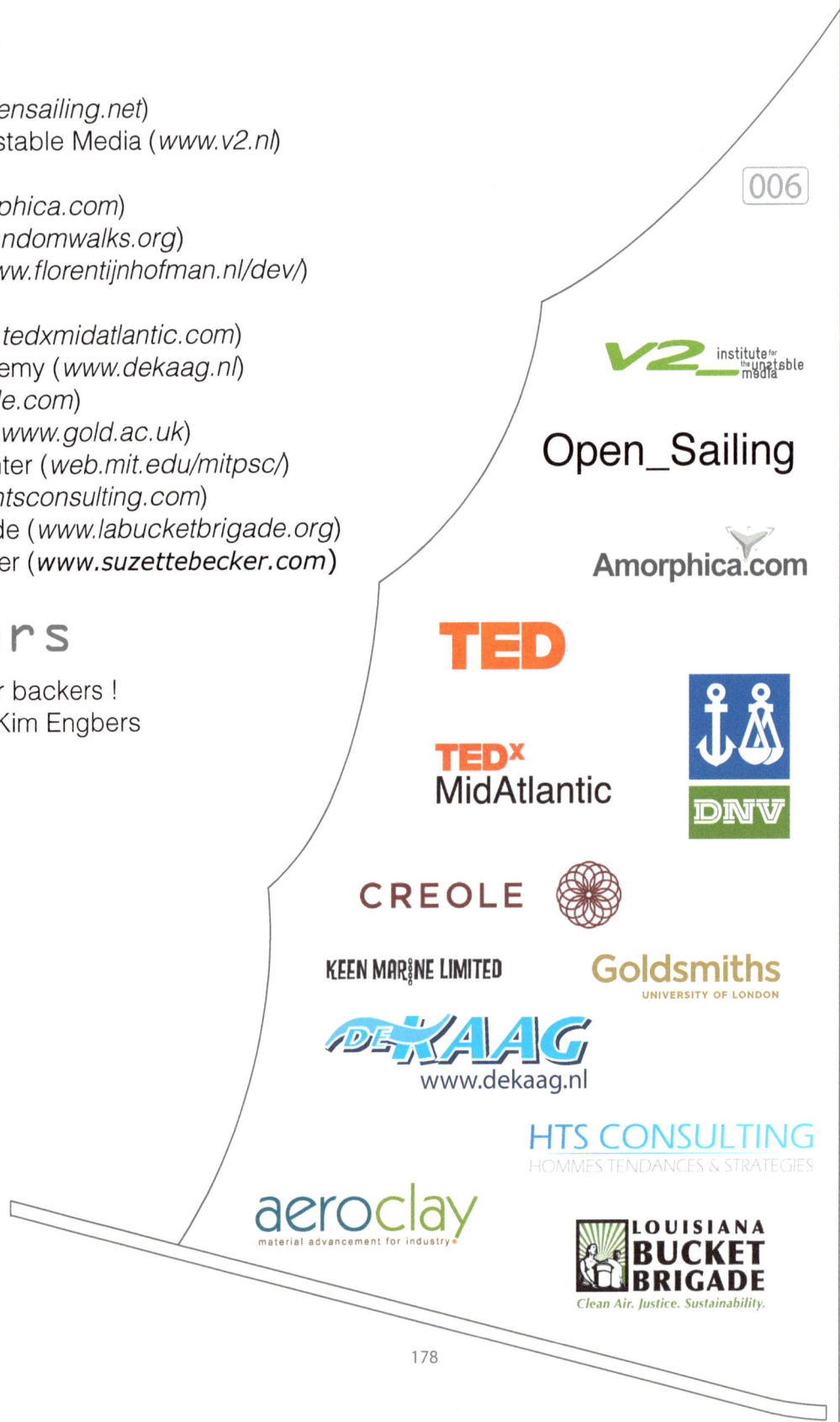

006

178

ø3.5

ø4.5

References

1. Kerr, Richard A. *A Lot of Oil on the Loose, Not So Much to Be*

2. *Found*. Science 329 (5993): 734–5. doi:10.1126/science.329.5993.734. PMID 20705818. "http://www.sciencemag.org/content/329/5993/734" 13 August 2010

3. http://www.slideshare.net/iaea/marine-environment-monitoring-of-fukushima-nuclear-accident-2-june-2011

4. Fogarty, David. *Navy to help climate scientists in pirate-infested waters*. Reuters online magazine. http://www.reuters.com/article/2011/07/14/us-climate-robots-idUSTRE76D16M20110714. 14 July 2011

5. Image Credit: Kangel http://commons.wikimedia.org/wiki/User:Kangel (Accessed 21 Aug 2011)

6. Fingas, Mervin F. *Oil Spill Science and Technology: Prevention*, Response, and Cleanup. Burlington, MA: Elsevier/Gulf Professional Pub., 2011. Print.

7. Fingas, Mervin F., and Jennifer Charles. *The Basics of Oil Spill Cleanup*. Boca Raton: CRC, 2001. Print.

8. ASTM F716 – 09, *Standard Test Methods for Sorbent Performance of Absorbents*

9. ASTM F726 – 06, *Standard Test Methods for Sorbent Performance of Adsorbents*

10. SINTEF *Oil Weathering and Appearance at Sea* – May 2011

11. IMO Circular letter No.1886/Rev.3, *IMO identification number scheme*

12. NOAA, *A Guide for Spill Response Planning in Marine Environments,* June 2010 (joint publication of U.S. DEPARTMENT OF COMMERCE, U.S. Coast Guard, U.S. Environmental Protection Agency, American Petroleum Institute)

13. OSPAR Recommendation 2001/1 for the Management of Produced Water from Offshore Installations, 2001

14. Aeroclay Inc. official website – http//aeroclayinc.com/ – Aug. 22nd 2011

15. US EPA – National Oil and Hazardous Substances Pollution Contingency Plan, 2011

16. Anne Rolfes, LA Bucket Brigade, Oil Spill Crisis Map Initial Analysis and Review, October 2010.

17. Extended Guidelines for the Use of the Bonn Agreement Oil Appearance Code (BAOAC), Annex 4, 2003

18. Worm B, Hilborn R, et al. (2009) Rebuilding Global Fisheries. Science

19. Surveillance and Enforcement of Remote Maritime Areas (SERMA), Marine Conservation Biology Institute

20. Agnew DJ, Pearce J, Pramod G, Peatman T, Watson R, et al. (2009) Estimating the Worldwide Extent of Illegal Fishing. PLoS ONE 4(2)

21. FAO (2000) The State of the World Fisheries and Aquaculture 2000. FAO, Rome, Italy

006

Protei
Open Source Sailing Drone

178

ø3.5

ø4.5

www.ingramcontent.com/pod-product-compliance
Lightning Source LLC
Chambersburg PA
CBHW041721210326
41598CB00007B/737